Petrochemicals 2023

Topical Conference at the 2023 AIChE Spring Meeting and 19th Global Congress on Process Safety

Houston, Texas, USA
12-16 March 2023

ISBN: 978-1-7138-7929-9

Printed from e-media with permission by:

Curran Associates, Inc.
57 Morehouse Lane
Red Hook, NY 12571

Some format issues inherent in the e-media version may also appear in this print version.

Printed with permission by Curran Associates, Inc. (2023)

For permission requests, please contact AIChE
at the address below.

AIChE
120 Wall Street, FL 23
New York, NY 10005-4020

Phone: (800) 242-4363
Fax: (203) 775-5177

www.aiche.org

Additional copies of this publication are available from:

Curran Associates, Inc.
57 Morehouse Lane
Red Hook, NY 12571 USA
Phone: 845-758-0400
Fax: 845-758-2633
Email: curran@proceedings.com
Web: www.proceedings.com

TABLE OF CONTENTS

PROCESS CONTROL AND OPTIMIZATION DEVELOPMENTS I

SUSTAINABILITY DEVELOPMENTS FOR REFINING AND PETROCHEMICALS I

DEVELOPMENTS IN PETROCHEMICALS

SUSTAINABILITY DEVELOPMENTS FOR REFINING AND PETROCHEMICALS II

PROCESS CONTROL AND OPTIMIZATION DEVELOPMENTS II

PROCESS CONTROL AND OPTIMIZATION DEVELOPMENTS III

REAL-TIME APPLICATIONS OF DATA ANALYTICS AND MACHINE LEARNING II

CATALYTIC ANSWER TO A STEAM CRACKING CHALLENGE: CS_2 CONTAMINATION

Michael Gaas[1], Marielle Gagniere[2], Esma Rokia[1], Yoeugourthen Hamlaoui[3] and Dr. Anne-Agathe Quoineaud[2]

1 Axens North America, 1800 Saint James Place, Houston, 77056, TX, USA

2 Axens North America, 650 College Road East, Princeton, 08540, NJ, USA

3 Axens, 89 Bd Franklin Roosevelt, Rueil Malmaison, 92500, France

Abstract

As Petrochemical production capacity has continued to expand globally, many contaminates have presented themselves as a new challenge to cracker production reliability. Olefins producers across the world, have begun to encountered a developing problem of carbon disulphide (CS_2) contamination by way of opportunity liquid feeds. This presence of CS_2 can be attributed to the use of additives upstream.

More and more, petrochemical producers are faced with the challenges, and repercussions, of high CS_2 content in the naphtha feedstock for steam crackers. In general, CS_2 has a strong impact on the selective hydrogenation catalyst activity, but this impact is especially true for Pygas hydrogenation units. This contamination yields the poisoning of Pygas catalyst, potentially resulting in off spec products, increased opex, early catalyst change outs, and unexpected outages/down time. As this issue presents itself to producers worldwide, Axens has dedicated itself to providing solutions today for the challenges of the future.

Axens newest, ultra-high activity, palladium based catalyst has been fully developed and tested with a singular goal; to tackle the challenge of CS_2 contamination. When comparing the performance of Axens Next-Gen LD 585 to industry leading referenced catalyst, Axens has found improved catalytic performance under high sulfur-containing feedstock's. The development of this next generation catalyst is thanks to Axens dedication to improving catalyst production and preparation.

Introduction

Ethylene is the most used monomer in the petrochemical industry, employed as feedstock in several manufactures. Over the past few years, global ethylene demand has grown rapidly. Asia, the Americas, and the Middle East have been leading new ethylene capacity addition in the past decade. For the upcoming decade, China plans to continue leading ethylene growth, with new steam crackers popping up around the country. North America has also invested heavily in new cracker projects in recent years due to an advantage in ethane feedstock supplied. Facing this unprecedented competition, current ethylene producers are exploring several options to improve their competitiveness:

- Revamping ethylene capacity,
- Improving refining and petrochemical integration,
- Using feedstock flexibility with diesel, crude, gas, or naphtha,
- Using new feedstock sources

Opportunity feedstocks offer a main advantage of lower cost. However, each advantage often has a counterpart; these feedstocks come with disadvantages. Among these disadvantages, one has been identified over the years as a strong contaminant for downstream catalysts, carbon disulphide (CS_2).

From well to steam cracking furnace

More and more petrochemical actors are facing high CS_2 content in the feedstock for their naphtha steam crackers, and this tendency is expected to increase in the coming years. Indeed, crude wells are ageing and, as a consequence, more H_2S and heavy materials like asphaltenes/ paraffins are present, making crude oil extraction more difficult. H_2S could be an issue for personnel safety in case of exposure during extraction or even during transportation. H_2S scavengers employed, such as triazine, amines, or formaldehydes, react with the carbon in crude oil to form complexes (see example in Figure 1).

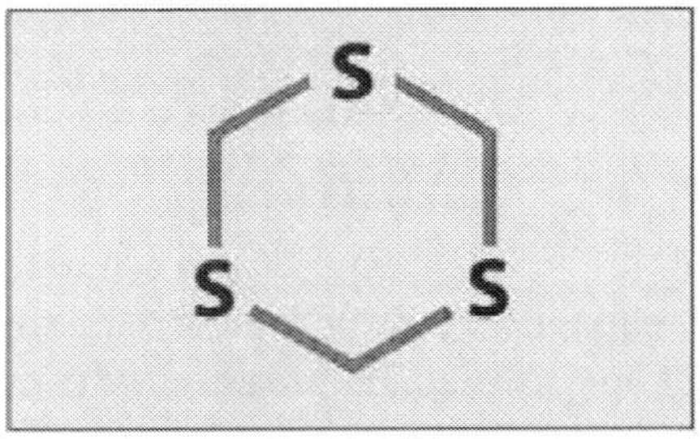

Figure 1. Example of Complex formed with carbon and CS_2

In the furnace, these complexes are cracked to form CS_2. "Heavy" hydrocarbons present problems for crude oil transportation or storage and, depending on their content, plugging can occur in the wellhead and in storage. CS_2 is then used as a flow improver and found in steam cracker furnaces. It would appear that CS_2 issues could be directly related to the naphtha source. Indeed, when importing naphtha from Brazil, Canada, Egypt, Kuwait, Morocco, Russia, and the US, operators often face this issue.

From cracking furnaces to downstream units

To understand where CS_2 impacts the steam cracker operation, it is worth bearing in mind the boiling point of CS_2 . This boiling point is close to that of some hydrocarbons with five carbon atoms (see Figure 2). Downstream of the stream cracker furnaces, several fractionation steps target recovery of C_2 (acetylene, ethylene, and ethane), C_3 (methylacetylene, propadiene, propylene, and propane), C_4 (butadiene, butene1, butene2, butane, and sometimes vinyl acetylene), and the C_5^+ cut (heavier components including aromatics as benzene and toluene). As a direct consequence, CS_2 is found in the C_5^+ cut.

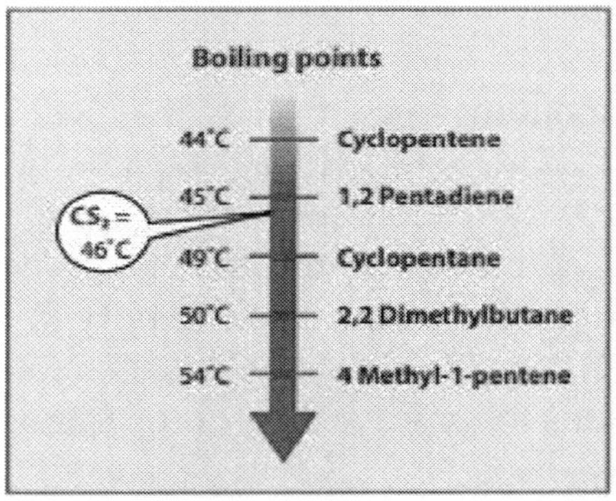

Figure 2. Boiling points of CS_2 and C_5 hydrocarbons

The C_5^+ cut may be used at several stages for different purposes: (i) The C_5 cut can be fully hydrogenated in a total C_5 hydrogenation unit to recycle the C_5 stream to the steam cracker furnaces and hence increase light olefins yield, (ii) C_5 can be used in an extraction unit to valorize the isoprene cut, and (iii) C_5^+ can be treated as a pyrolysis gasoline (pygas) cut in a pygas hydrogenation unit to allow aromatics (benzene, toluene) recovery while removing the sulfur component. Between the steam cracking furnaces and the C_5^+ cut, the CS_2 content becomes concentrated. Therefore, 4-5 wtppm of CS_2 in the naphtha feedstock will lead to 20-25 wtppm of CS_2 in the pygas feed, considering typical pygas yield on the naphtha cracker (see Figure 3). This explains how CS_2 has been identified by comparing the lack of catalyst activity observed in C_5 total hydrogenation and pygas hydrogenation with the origin/nature of the naphtha feedstock.

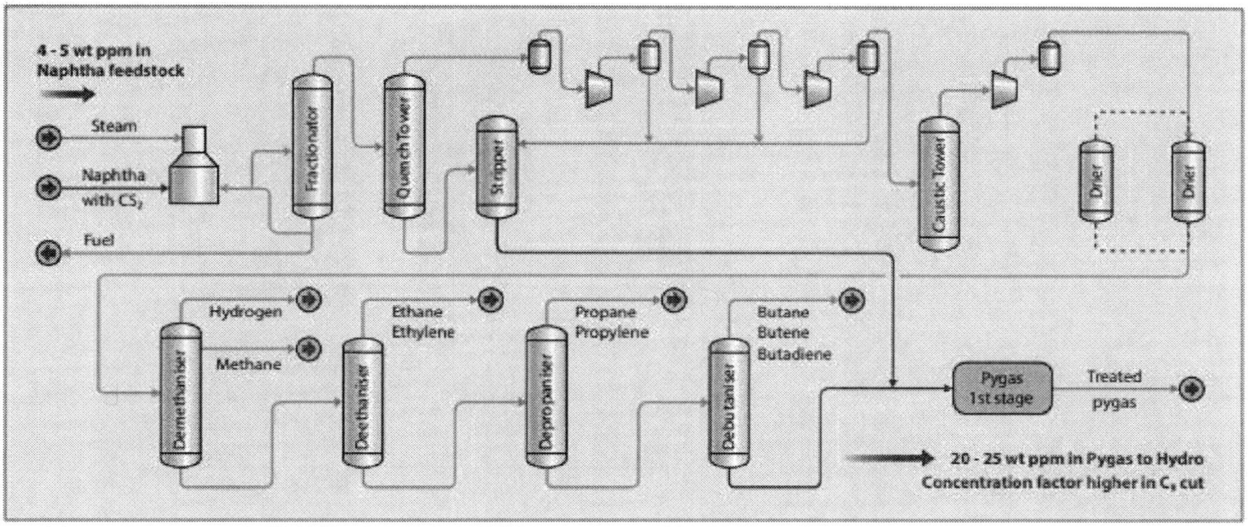

Figure 3. CS_2 concentration in the C_5^+ cut

CS2 impacts

Pygas feed may contain up to several hundred parts per million of sulfur. Speciation of sulfur, including CS_2, has highlighted the presence of thiophenes (80 wt% of the total sulfur species), mercaptans/sulfides/disulfides (15 wt%), CS_2 (5 wt%), and H_2S/COS (below 0.5 wt%). These sulfur species have different poisoning effects on the catalyst used in C_5/pygas hydrogenation. Among the sulfur species described here, thiophenes present the lowest poisoning effect, followed by mercaptans, sulfides, and disulfides in ascending order. H_2S/COS has the strongest poisoning effect and CS_2 the second strongest poisoning effect. The strong poisoning effect of CS_2 has been observed in industrial units and confirmed by pilot tests. Indeed, in the presence of a few ppm of CS_2 in the pygas feed, a reactor temperature increase is required to compensate the catalyst deactivation.

A pygas unit comprises two stages. The first stage targets complete removal of diolefin and styrenic components while the second stage aims at the complete removal of sulfur and olefin components while ensuring lower aromatics losses. Figure 4 shows industrial feedback from a first stage pygas unit operated with a palladium based catalyst. This unit encounters CS_2 inhibition on a regular basis. To the left, the position of the thermocouple is indicated in the catalyst beds. The reactor exotherm distribution per thermocouple is shown to the right. This exotherm distribution provides an accurate picture of catalyst activity. Indeed, if the catalyst is not inhibited, the exotherm should be mainly present at the top part of the catalyst bed. If contamination is present in the feed, we observe catalyst activity shifting from the top part of the catalyst bed to the bottom part.

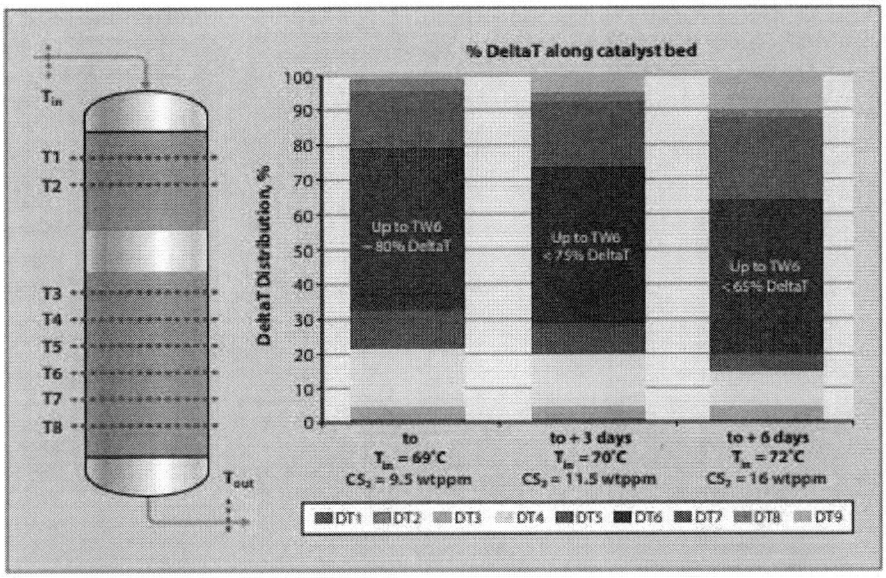

Figure 4. CS_2 impact on pygas palladium based catalyst

For the operational history summarized here, 9.5 wtppm of CS_2 is considered to be the initial instant T0, and the required styrene conversion is achieved with a 69°C of reactor inlet temperature (RIT). The CS_2 level increases over the days of operation. At T0 + 6 days, the CS_2 content is around 16 wtppm; however the styrene content is multiplied by more than 1.5 in the product, even with a 3°C increase in RIT. Figure 4 shows that even a few parts per million of CS_2 have a strong impact on catalyst activity. Nevertheless, the product is brought back on-specification once CS_2 is no longer present in the feed. This observation confirms that CS_2 acts as a strong inhibitor for the palladium based catalyst. This industrial scale observation is also backed by pilot test results.

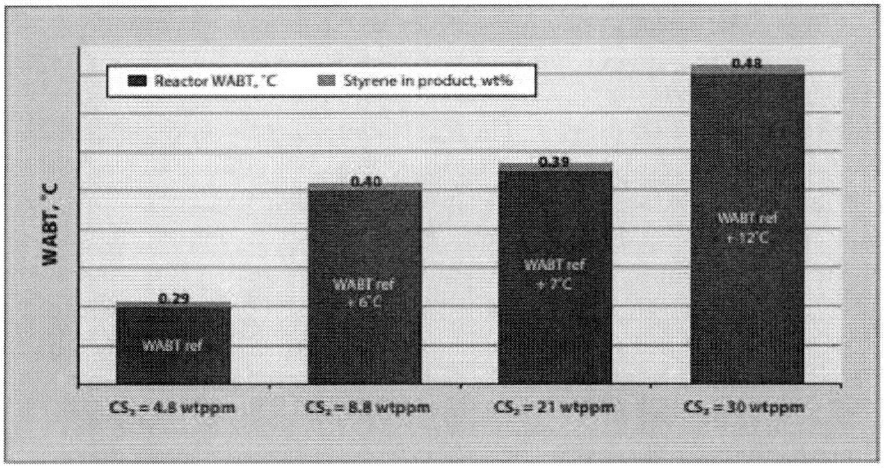

Figure 5. CS_2 impact on pygas nickel based catalyst

Figure 5 shows the feedback from a first stage pygas unit operating with a nickel based catalyst. This unit also encounters CS_2 inhibition on a regular basis. The weight average bed temperature (WABT) and styrene in the product are shown for several values of CS_2 in the feed at a constant operating flow rate. The operating point of reference corresponds to a product styrene content of around 0.3 wt% with 5.0 wtppm of CS_2. From the initial point to 9.0 wtppm of CS_2, WABT is increased by 6°C to achieve 0.4 wt% of styrene in the product. From 9.0 wtppm to 20 wtppm of CS_2, the nickel catalyst exhibits almost no deactivation. At 30 ppm of CS_2 in the feed, the WABT needs to be increased by 12°C more than the reference WABT while the product styrene is around 0.5 wt%, which is above the maximum target content. Industrial feedback shows that activity is not recovered when CS_2 is no longer present in the feed. The catalyst is poisoned and a specific catalyst treatment is required to recover its activity. A clear conclusion emerges when the various observations are combined. A nickel based catalyst, although it needs less temperature compensation than palladium to resist CS_2 contamination within the range 10-20 wtppm, nevertheless tends to be poisoned by a high CS_2 level. However, a palladium based catalyst is more active than a nickel catalyst and recovers its activity more easily without any specific treatment. Considering the relative activity of palladium compared to nickel along with the observations reported here, Axens decided to focus first on the development of a specific palladium catalyst displaying enough activity to ensure that it achieves the required performance even at high content of CS_2 in the feed. Axens continues to develop the most suitable solution for units currently using nickel based catalyst.

CS2 contamination: solutions

For isoprene extraction, Axens can propose a commercially demonstrated, adsorbent based solution which selectively traps CS_2 from the isoprene cut. The adsorbent efficiency is boosted by a feed composition containing few types of sulfur species.

In C_5 hydrogenation and pygas hydrogenation, the challenge is quite different. As we have seen, the feed contains a wide variety of sulfur species. The solution should also be compatible with the catalyst system. It was decided to develop a new catalyst displaying the highest level of activity to ensure meeting the required level of styrenics/ diolefins hydrogenation despite CS_2 contamination. Once the first stage of pygas duty was achieved, the second stage would then be efficient enough to remove CS_2 along with the other sulfur species from the pygas feed.

A new palladium catalyst was developed focusing on this need for higher activity. In order to achieve this target, the catalyst (LD 585) has a very open pore structure under an extrudate shape combined with a fine distribution of palladium to maximize the accessibility of active palladium and thus increase significantly the catalyst's activity.

Pilot tests were performed to assess LD 585's absolute level of activity and compare it to existing catalytic solutions. Some of the pilot tests performed to validate LD 585's performance are described here.

Experiments

Unit description

A three-reactor isothermal plug flow pilot test unit was used to compare the performance of different catalysts. The reactors are in upflow mode, double piped, with external high velocity thermal fluid used to guarantee a flat temperature axial profile in the catalyst bed. Thermocouples are used in the catalyst bed to adjust the reaction temperature. Gas and effluent are sent separately to the same gas chromatography system. An automated control system is used to monitor the sequences of test and analysis.

Testing protocol

Raw pygas feed rather than synthetic feed was used for this test, to adhere to industrial constraints in terms of styrene content and diene value. The feed drums were blanketed under nitrogen. Feed quality was controlled throughout the test's duration as the pygas feed is known to be prone to ageing which leads to conversion of cyclopentadiene to dicyclopendatiene. The sulfur level was low in the raw pygas feed used for the test, with no CS_2 present. This allowed more precise control of the CS_2 content and its effect on catalyst activity by spiking CS_2 in the raw feed.

Test methodology

One mass balance was carried out per reactor per operating condition at the end of each plateau. The effluent of each reactor is in mixed phase. The total mass balance is determined by comparing the reactor inlet (liquid phase) with a recombination of the reactor outlet (mixed phase). The liquid mass balance is determined by weight measurement and the gas mass balance by dry gas flow meter. Styrene activity and conversion are continuously followed up with gas chromatography analysis.

Catalyst characteristics

In each reactor, 15 cc of Axens catalysts are loaded: a high activity nickel based catalyst (a leading Ni catalyst), a high activity palladium based catalyst (a leading Pd catalyst), the new high activity palladium based catalyst (LD 585). The objective of the test was to compare the activity level of each catalyst proportionally to CS_2 content in order to maintain 90% styrene conversion. Taking into account the relative activity of each type of catalyst, both palladium and nickel, the liquid hourly space velocity (LHSV) is identically set up in order to mimic faithfully the industrial operating conditions for the same reactor volume. Indeed, LHSV is usually higher for a palladium type than for a nickel based catalyst due to the higher activity of palladium catalyst. The temperature is controlled and adjusted to compensate catalyst deactivation following variation of the CS_2 content during the tests.

Results

The test, carried out over 87 days, follows four successive sequences (see Figure 6). The sequence 1 corresponds to a low sulfur raw pygas feed content. Catalysts were tested with raw pygas at 50 ppm of sulfur in the feed (typical sulfur species from pygas feed – no CS_2). The test required 300 hours of operation to adjust the operating parameters. Under raw feedstock stabilized conditions, the three catalysts displayed different levels of activity. The most active catalyst is the newest catalyst, LD 585, with the leading palladium catalyst second and the leading nickel catalyst clearly lower in activity than both palladium based catalysts, with conversion, respectively, of 99%, 80%, and 10 %. The leading nickel catalyst, the leading palladium catalyst, and LD 585 were then settled in temperature to achieve the same styrene conversion before Sequence 2 began.

The sequence 2 corresponds to a high sulfur raw pygas feed content. Catalysts were tested with raw pygas spiked with 50 ppm of CS_2 for 100 ppm of total sulfur. A sudden deactivation of the leading palladium catalyst and the leading nickel catalyst was observed. LD 585 maintained a good level of activity, with 50% styrene conversion instead of 100%, while the leading palladium catalyst could achieve only 20% styrene conversion and the leading Ni catalyst achieved only a few percent styrene conversion. The temperature was adjusted in order to reach 90% styrene conversion for all of the catalysts. Then, the reactors were maintained at these operating conditions in order to test the catalysts' stability regarding deactivation up to 1300 hours. LD 585 (red spheres in Figure 7), the leading palladium catalyst (blue triangles), and the leading nickel catalyst (green squares) were quite stable at this sulfur content with a loss of activity of about 20%. Under these operating conditions, the leading palladium catalyst and the leading nickel catalyst showed good stability while LD 585 had the lowest deactivation rate with almost no decrease in activity. Based on the deactivation rates observed during Sequence 2, it was decided to greatly increase the CS_2 content in the feed.

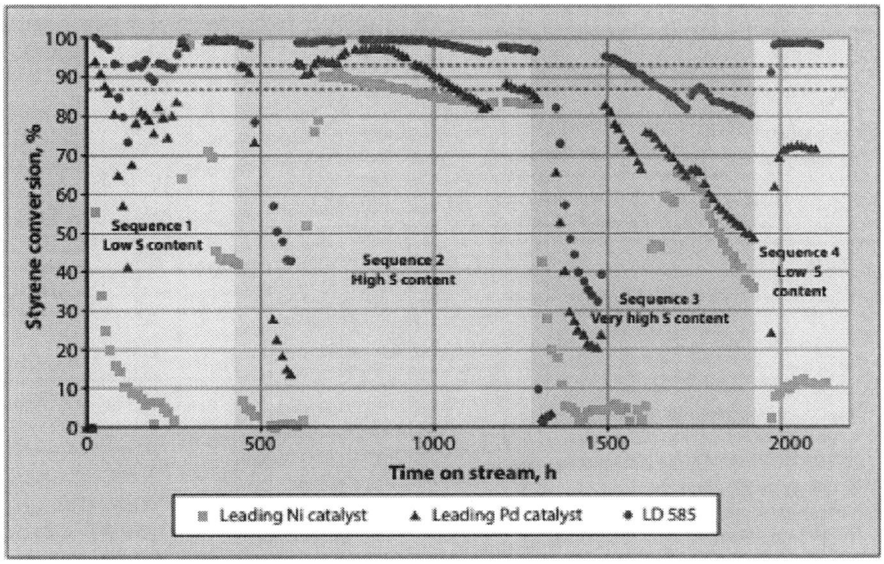

Figure 6. Test results at different CS_2 contents with raw pygas feed

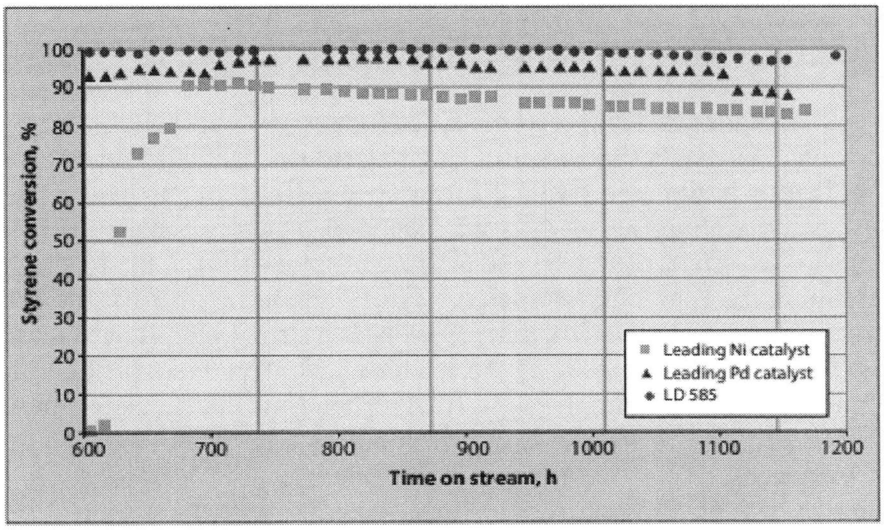

Figure 7. 26-day test at 50 ppm CS_2

The sequence 3 corresponds to a very high sulfur raw pygas feed content. The catalysts were tested with raw pygas spiked with 350 ppm of CS_2 for 400 ppm of total sulfur. CS_2 was injected into the three reactors at the same time (1400 hours). All three catalysts showed strong and sudden deactivation (see Figure 8). For LD 585, styrene conversion was 32% while the leading palladium catalyst's styrene conversion was only 20%. For the leading nickel catalyst, it was only 8%. The temperature was increased to achieve the highest possible styrene conversion. For LD 585, the reactor temperature was set at 150°C, which led to 95% styrene conversion (see Figure 8). With the leading palladium catalyst, the temperature was adjusted at 155°C to obtain 70% conversion. For the leading nickel catalyst, the temperature was increased to 180°C (the maximum allowable temperature in first stage pygas), with only 65% styrene conversion achieved; this was followed by strong deactivation. At this level of operating temperature, several phenomena can occur, for instance sulfur penetration into the heart of the nickel particles. The formation of Ni_3S_2 species, which are inactive in hydrogenation catalysis, can explain this decrease in conversion and activity. The three reactors were left at these operating conditions. While the leading palladium catalyst and the leading nickel catalyst showed almost 60% loss of activity with a sharp rate of deactivation, LD 585 demonstrated stability and maintained a high level of performance even with 350 ppm CS_2 in the feed.

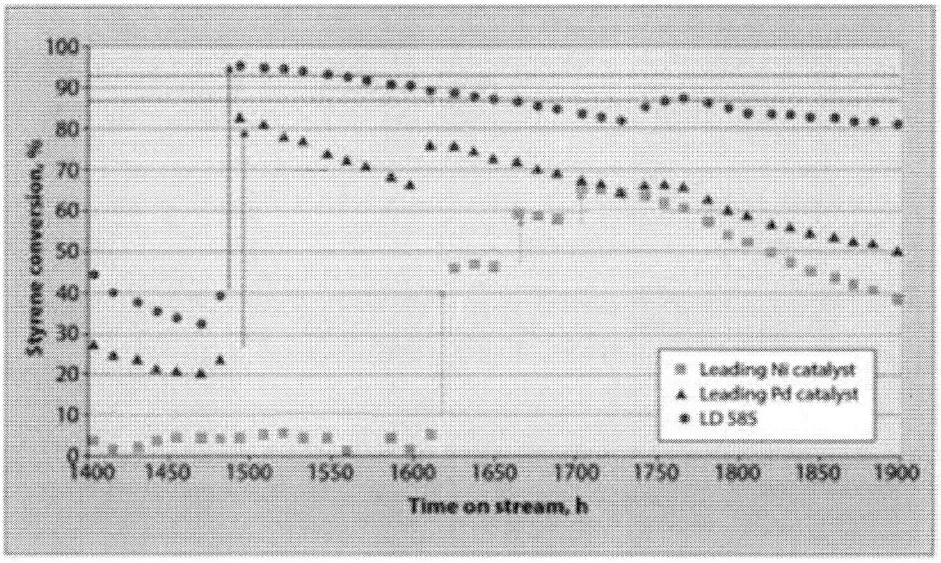

Figure 8. Temperature adjustment at 350 ppm of CS_2

The sequence 4 corresponds to low sulfur raw pygas feed content. The catalysts were tested with raw pygas at 50 ppm sulfur in the feed (typical sulfur species from pygas feed – no CS_2) to determine recovery of catalyst activity. Spiking of CS_2 was stopped at approximatively 2000 hours. This last sequence corresponds to Sequence 1 (no CS_2 and 50 ppm of total sulfur in the feed). The aims of this step are two-fold: to determine the capability of each tested catalyst to recover its initial activity once CS_2 is no longer present in the feed, and to confirm that CS_2 acts as a strong inhibitor. With feed conditions back to their initial level, the remaining activity of the leading nickel catalyst was quite low, achieving only 10% styrene conversion. Compared with the activity of the leading nickel catalyst during Sequence 1, this corresponds to approximatively 1% of conversion loss per day, in other words 80% of total conversion loss during the test sequences.

The nickel based catalyst did not recover its initial activity without special treatment once it had operated with feed containing a high amount of CS_2 . The leading palladium catalyst achieved 70% conversion during Sequence 4 with the same adjusted operating temperature as in Sequence 3. The leading palladium catalyst recovered part of its activity once CS_2 was no longer present in the feed, however catalyst deactivation here is quite important over time. The leading palladium catalyst was in service for 87 days with high CS_2 fluctuations, which explains a total loss of activity of around 70% compared with the activity it displayed at the beginning of the test. By contrast, LD 585 lost only 10% of its activity during the entire test. Showing simultaneously high CS_2 resistance, excellent stability, and activity recovery, LD 585 has demonstrated its efficiency in handling the CS_2 issue through its ultra-high intrinsic activity.

Conclusion

Axens discovered the CS_2 issue in 2015 in discussions with a customer facing unexplained loss of activity in a pygas first stage unit. This would become a worldwide challenge for the steam cracker environment. A lear profile of this new contamination emerged from joint work between Axens and the customer. CS_2 acts as a strong inhibitor and arrives with naphtha feed from regions where crude oil is becoming more and more difficult to extract. Once in the naphtha, CS_2 follows a clear track from the cracking furnaces to the C_5 cut and the pygas. Once the origin and the nature of the contamination are identified, the most difficult part begins: determining the best way to mitigate this serious contamination. Axens worked on several solutions, focusing on removing CS2 from the naphtha feed or developing a new catalyst displaying an ultrahigh level of activity to compensate for the inhibiting effects of CS_2 . LD 585 palladium based catalyst's high activity and stability are the first part of the solution developed by Axens to address this worldwide challenge.

MODIFICATION OF ZEOLITE-Y CATALYST BY ADDITION OF SELECTED TRANSITION METAL IONS (Zn &Cu ions)

[1]Patrick O. Nwosibe, [1]Okwara Akandu Uche, [1]Ibrahim Idris
[1]Department of Chemical Engineering, Federal Polytechnic, Kaduna
[2]Department of Civil Engineering, Federal Polytechnic,Kaduna

Email: patnwosibe@gmail.com, +2348037710269

ABSTRACT

Zeolite is a microporous nanomaterial with alumina and silicate structure which can be used to crack heavy petroleum fraction to yield lighter components. Recent studies suggests that zeolite can be used for sugar conversion to fuel and value-added chemicals with proper modification. In this study, fresh zeolite -y catalyst was subjected to initial characterization (BET, XRD, XRF & SEM) and modified. 50ml of prepared standard solution of zinc and copper chloride each was added to 5g of fresh zeolite catalyst and mixed thoroughly for 1 hour at room temperature. The mixture was dried at 110 ^0C for 24 hours using dry oven. The crystalline solid obtained was calcined at 500 ^0C for 2 hours in a muffle furnace. The modified catalyst was finally characterized (BET, XRD, XRF and SEM) and the results obtained was compared with the fresh catalyst characterized results. BET results shows that the pore volume, pore diameter and surface area of the modified catalyst increase from 0.309cc/g, 2.105nm, and 627.998m2/g to 0.365cc/g, 2.446nm and 746.062m2/g respectively. Similarly, from the XRD result, the modified catalyst has all the crystalline peaks at Bragg angle (2θ) present in the fresh catalyst with additional peaks formed. The additional crystalline peaks formed shows the increase in mechanical strength of the modified catalyst. The SEM results shows the change in surface texture of the fresh catalyst from smooth and uniform to rough and agglomerated.

Keywords: Zeolite, catalyst, transition metals. modification, crude, ions

1.0 Introduction

Catalyst is any substance that increases the rate of a reaction without itself being consumed. Most solid catalyst are metals or the oxides, sulfides, and halides of metallic elements and of semi-metallic elements, boron, aluminum, and silicon. Gaseous and liquid catalyst are commonly used in their pure form or in combination with suitable carriers or solvents. In general, catalytic action is a chemical reaction between the catalyst and a reactant, forming chemical intermediates that are able to react more readily with each other or with another reactant to form the desired end product. During the reaction of the chemical intermediates and the other reactants, the catalyst is regenerated.

Fluid Catalytic Cracking (FCC) is the conversion process used in petroleum refineries to convert the high-boiling point, high-molecular weight hydrocarbon fractions of petroleum (crude oils) into gasoline, olefin gases, and other petroleum products. The cracking of petroleum hydrocarbons was originally done by thermal cracking, now almost replaced by catalytic cracking, which yields greater volumes of high -octane rating gasoline; and produces by-product gases, with more carbon double bonds (i.e. olefins), that are of greater economic value than the gases produced by thermal cracking.

The feedstock to the FCC conversion process usually is heavy gas oil (HGO), which is that portion of the petroleum (crude oil) that has an initial boiling-point temperature of 340 °C (644 °F) or higher, at atmospheric pressure, and that has an average molecular weight that ranges from about 200 to 600 or higher; heavy gas oil also is known as "heavy vacuum gas oil" (HVGO). In the fluid catalytic cracking process, the HGO feedstock is heated to a high temperature and to a moderate pressure, and then is placed in contact with a hot, powdered catalyst, which breaks the long-chain molecules of the high-boiling-point hydrocarbon liquids into short-chain molecules, which then are collected as a vapor.

Zeolites represent a revolution of crystalline porous materials. They are an important group of heterogeneous catalysts with large-scale applications in the petroleum industry and an increasing application potential in environmental catalysis. Zeolites are crystalline alumina silicates with 3D four-connected frameworks built by corner sharing SiO_4 and AlO_4 units and a Si/Al ratio greater than one. Other atoms with tetrahedral coordination such as Ge, B, and Ti can also be introduced into the framework and enrich the zeolite family (Foster et al., 2012).

2.0 Methodology

Materials

The following materials and reagents were used for this research; fresh zeolite catalyst, Zinc Chloride, Copper II chloride, Hydrochloride acid and distilled water. The materials were sourced from school laboratory and open market.

The equipment used for the research includes weighing balance, magnetic stiller, oven ,beakers, muffle furnace, measuring cylinder.

Methodology

Fresh catalyst (Zeolite Y) was obtained from Kaduna Refining and Petrochemical Company (KRPC), Kaduna state in Nigeria. The zeolite catalyst was subjected to initial characterization (**SEM, BET, XRD** and **XRF**) to ascertain its original nature in Ahmadu Bello University (ABU) multi-user laboratory before the modification process was carried out in the laboratory .

Preparation of Standard Solution of Zinc Chloride

13.63g of granulated zinc chloride was dissolved in a minimum amount of 0.1M (100 ml) HCL. 700 ml of purified water (distilled water) was added, mixed and allow to cool to room temperature. The volume was made up to 1000 ml with purified water. The solution was mixed thoroughly.

Preparation of Standard Solution of Copper Chloride

1.965g of cupric chloride accurately weighed, was dissolved in sufficient 0.1M (100 ml) of HCL. The volume was made up to 1000 ml with distilled water and mixed thoroughly.

Mixing

5g of the zeolite catalyst was introduced into a 500 ml beaker. 50 ml of zinc chloride and 50 ml of cupper chloride standard solution were added to the zeolite in the beaker. The solution was subjected to magnetic stirrer for mixing in order to achieve homogeneity. The stirring was done for 2 hours at room temperature (25 °C)

Drying

In order to remove the liquid solution, present in the mixture, the mixture was dried to a crystalline solid in a dry oven at 110 °C for 24 hours.

Calcination

The dried crystalline zeolite was calcined in a muffle furnace at a temperature of 500 °C for 2 hours. Finally, the calcined zeolite was characterized (i.e. **SEM, BET, XRD and XRF).**

3.0 Discussion of Results

XRD Analysis:

Quantitative analysis report

General information

Analysis date	2022-03-23 10:39:00	Measurement start time	2022-03-23 09:58:47
Analyst	Administrator	Operator	Administrator
Sample name	ZEOLITE	Comment	
Measured data name	C:\WallPaper\23-03-2022\ZEOLITE_20220323_095847_G01_S...	Memo	

Multiple Profile

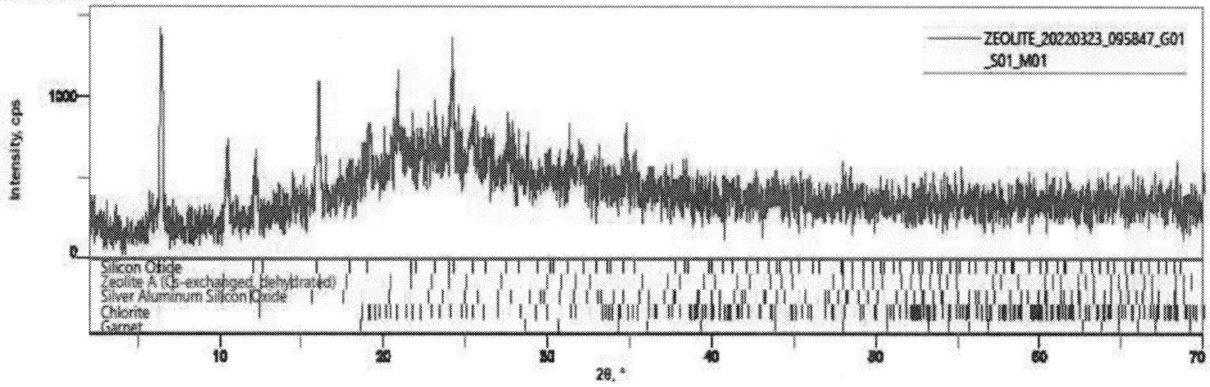

Path: C:\WallPaper\05-06-2022\ZEOLITE A Solution.mrsln Solution

Quantitative analysis report

General information

Analysis date	2022-07-05 13:44:51	Measurement start time	2022-07-05 12:42:08
Analyst	Administrator	Operator	Administrator
Sample name	ZEOLITE A	Comment	
Measured data name	C:\WallPaper\05-06-2022\ZEOLITE A_20220705_124054_G01...	Memo	

Multiple Profile

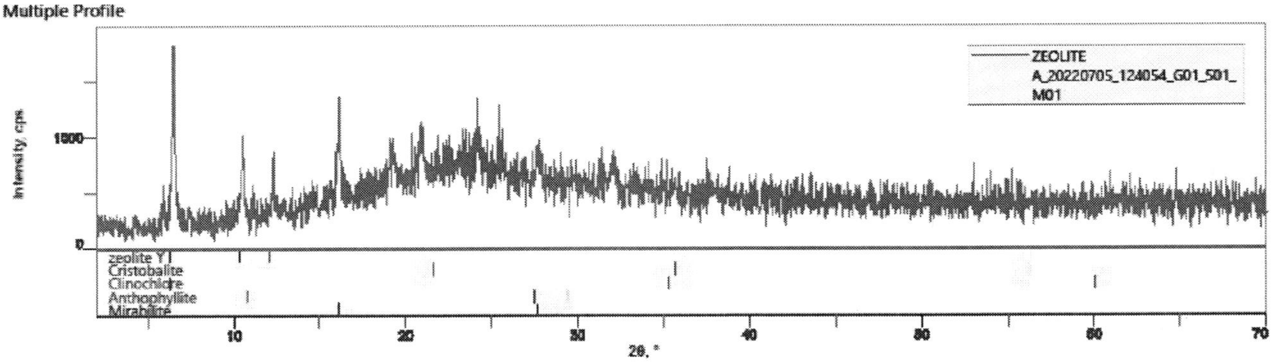

Quantitative analysis report (Solution) Page 1 of 2

From Figure 4.1 above, the XRD pattern shows that, the crystallinity of the fresh catalyst at 2θ = 7°, 10.2°, 12°, 15°, 22°, 25°, 31° and 35°. From FIG. (4.2), the modified catalyst has all the crystalline peaks present in the fresh catalyst with additional peaks formed. The additional peaks formed are; 2θ = 16°, 19°, 27°, 28.5° and 33°. this implies that the more active crystalline peaks are formed, the more the mechanical strength or activity increases and vice versa. This follows the trend as reported by Nwosibe *et al.* (2018).

XRF Analysis:

XRF Results for Test Sample of Fresh and Modified Catalyst.

Oxides	**Fresh Catalyst**		**Modified Catalyst**	
	Concentration	**Mole (%)**	**Concentration**	**Mole (%)**
SiO_2	**44.226**	**54.398**	**51.310**	**61.60**
Al_2O_3	**32.996**	**23.347**	**41.303**	**29.938**
ZnO	**0.011**	**0.010**	**6.425**	**5.695**
TiO_2	**2.700**	**2.498**	**3.397**	**3.069**
Cl	**0.608**	**1.267**	**2.064**	**4.201**
Fe_2O_3	**1.073**	**0.498**	**1.503**	**0.679**
CuO	**0.038**	**0.035**	**0.614**	**0.557**
V_2O_5	**0.358**	**0.145**	**0.358**	**0.142**
Cr_2O_3	**0.185**	**0.090**	**0.266**	**0.126**
NiO	**0.023**	**0.023**	**0.062**	**0.060**
CaO	**0.090**	**0.118**	**0.152**	**0.196**

From Table 3.1 above, the concentration and mole percent of the key oxides such as; SiO_2, Al_2O_3, ZnO, etc. decreased in the fresh catalyst. It can be seen from the above that, in the modified catalyst, the oxides increased in both concentration and weight percent. This is in line with what was reported by Foster et al. (2012).

BET Analysis:

Table 3.2: BET Surface Area, Pore Volume and Pore Diameter for Fresh and Modified Zeolite Catalyst.

Sample	Surface Area (M2/G)	Pore Volume (Cc/G Or Cm3/G)	Pore Diameter (Nm)
Fresh catalyst	**627.998**	**0.309**	**2.105**
Modified catalyst	**746.062**	**0.365**	**2.446**

The surface area, pore volume and pore diameter of the fresh catalyst were determined by BJH Adsorption Method and the results were summarized in Table 3.2 above. As shown in Table 3.2, the surface area, pore volume and pore diameter of the modified catalyst increased after the addition of the additives, especially zinc and copper chloride. This follows the trend as reported by Paasikollio et al. (2014).

SEM Analysis:

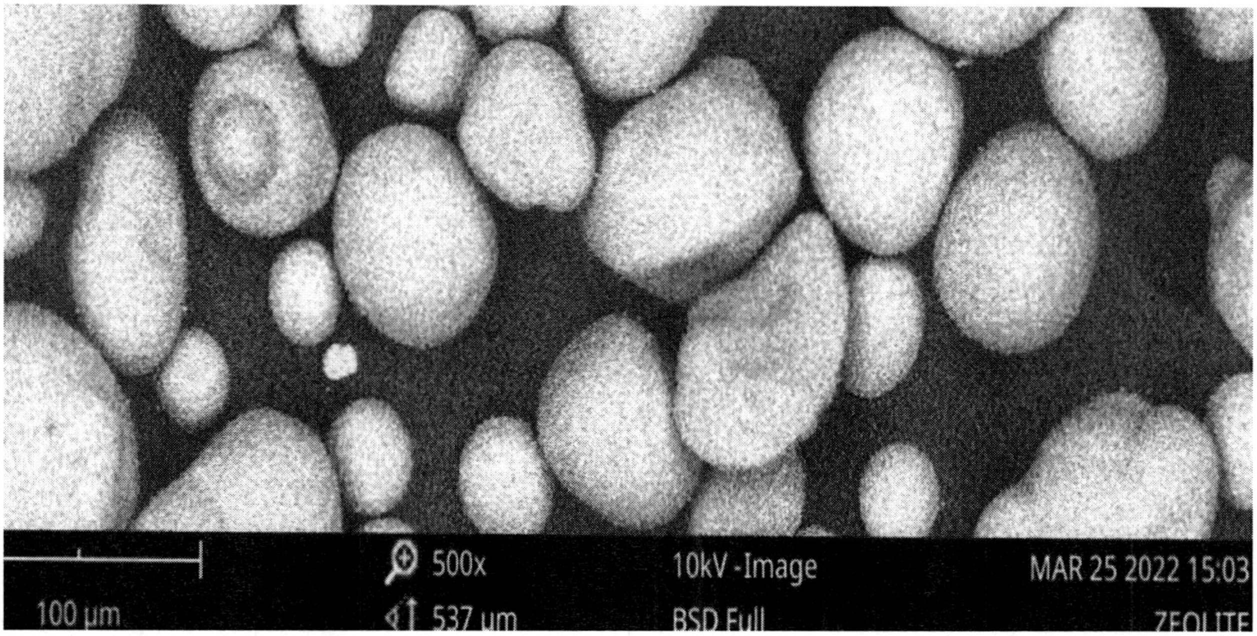
100 µm
500x
537 µm
10kV - Image
BSD Full
MAR 25 2022 15:03
ZEOLITE

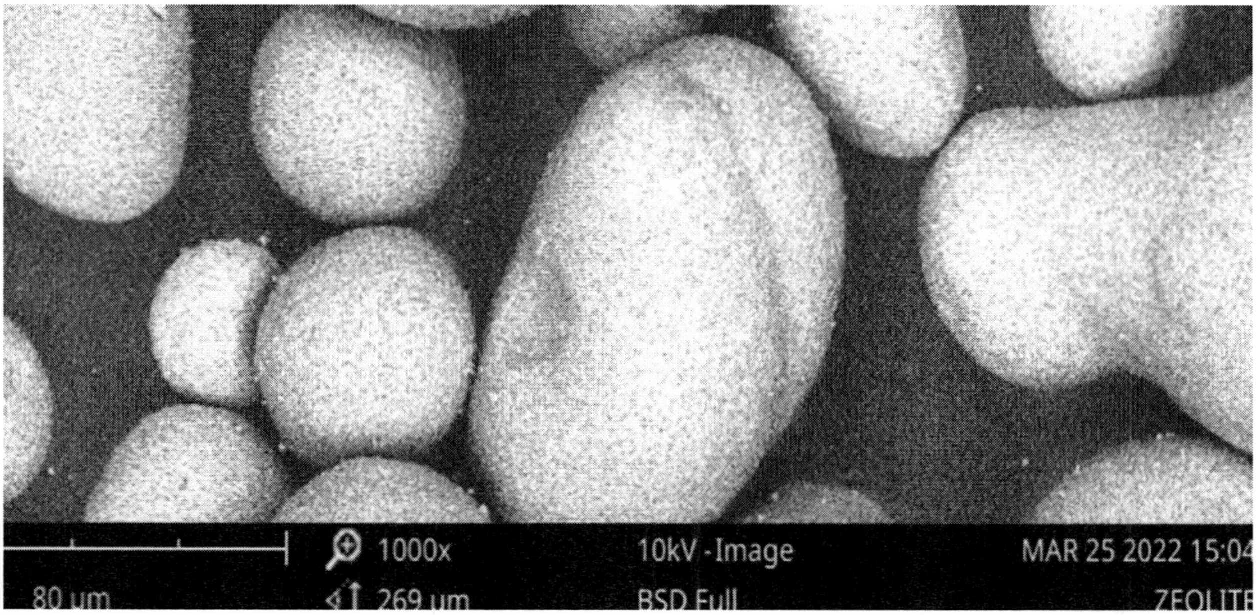
80 µm
1000x
269 µm
10kV - Image
BSD Full
MAR 25 2022 15:04
ZEOLITE

Figure 3: SEM Images of Fresh Catalyst

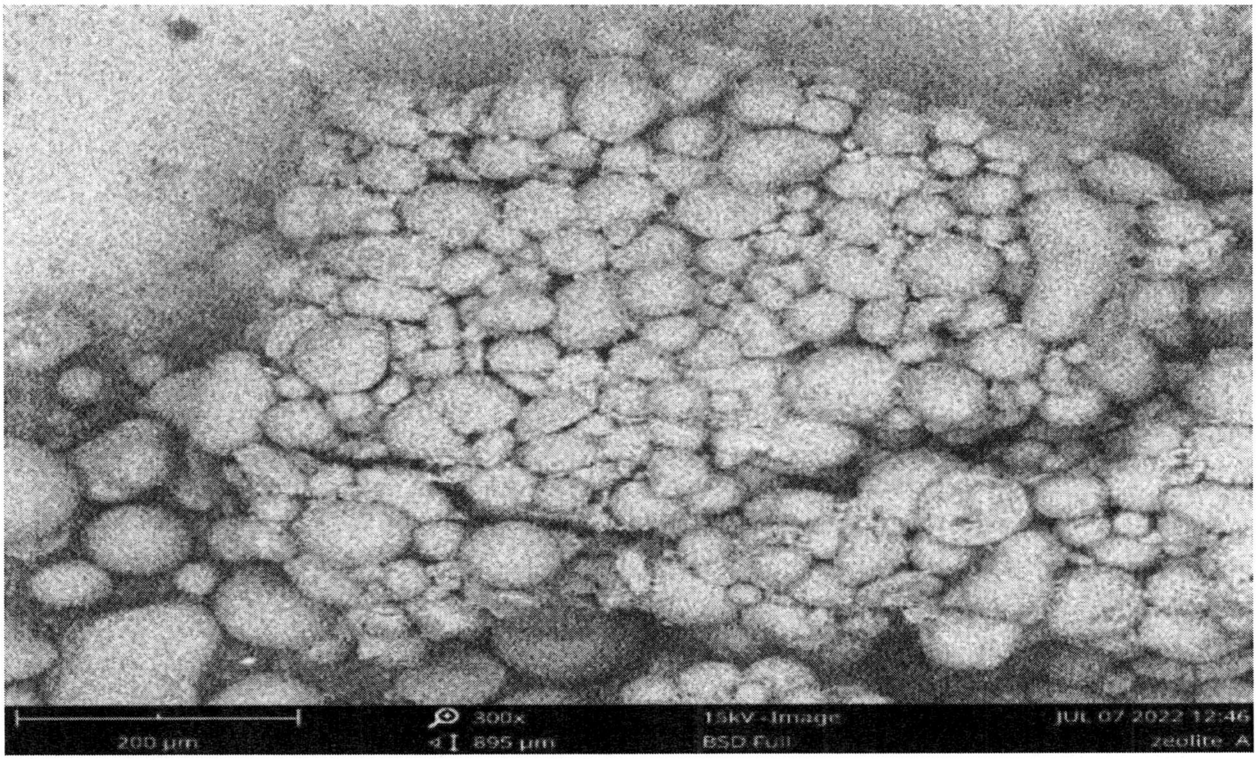
200 µm
300x
895 µm
15kV - Image
BSD Full
JUL 07 2022 12:46
zeolite_A

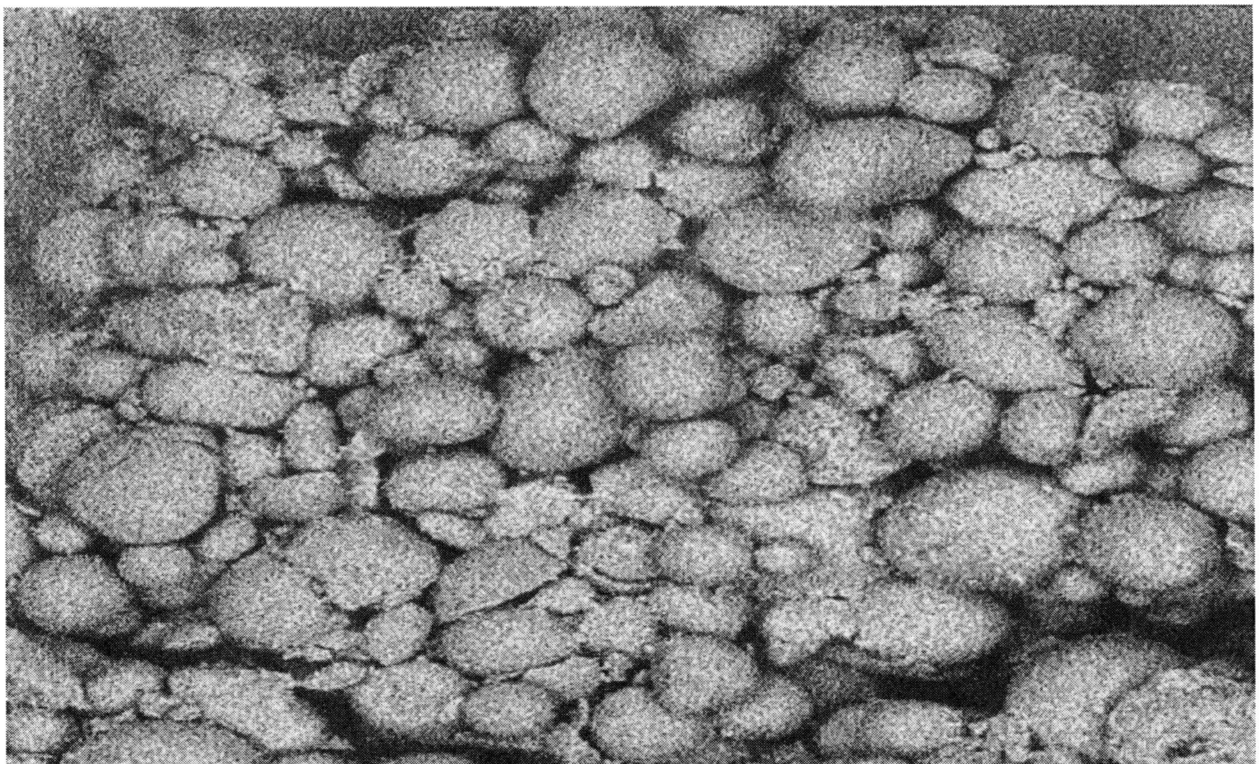
100 µm
500x
537 µm
15kV - Image
BSD Full
JUL 07 2022 12:46
zeolite_A

1000x
15kV - Image
JUL 07 2022 12:46
80 µm
269 µm
BSD Full
zeolite A

Figure 3: SEM Images of Modified Catalyst.

Scanning Electron Microscopy (SEM):

In order to observe the changes of the surface morphology and the amount of the surface elements of the fresh and modified catalyst, SEM experiment was conducted and the results are shown in figure 4.3 and 4.4 respectively. As illustrated in figure 4.3, the surfaces of the fresh catalyst are smooth and uniform. While after the modification with ZnO and CuO in figure 4.4, a significant agglomeration and sulfates deposited on the surface could be observed. This implies that the catalyst modified by ZnO and CuO could reduce the deposition of ammonium sulfate on the surface of the catalyst. This is in line with what was reported by Jin et al. (2014).

4.0 Conclusion

The fresh (commercial) zeolite was sourced and characterized. **XRD** pattern shows the major and minor crystalline peaks at Bragg (2θ). The concentration and weight % of the major oxides (SiO_2 and Al_2O_3) were determined from the **XRF** analysis. **BET** result shows the pore volume, pore diameter, and surface area of the fresh catalyst which were recorded in Table 4.2. It was observed from the **SEM** analysis conducted, the surface texture of the fresh catalyst was found smooth and uniform.

The fresh catalyst was modified by addition of a prepared standard solution of Copper and Zinc Chloride.

The modified catalyst was subjected to **BET, SEM, XRD** and **XRF** analysis. **BET** results shows that the pore volume, pore diameter and surface area of the modified catalyst increase from 0.309cc/g, 2.105nm, and 627.998m^2/g to 0.365cc/g, 2.446nm and 746.062m^2/g respectively. Similarly, from the XRD result, the modified catalyst has all the crystalline peaks at Bragg angle (2θ) present in the fresh catalyst with additional peaks formed. The additional crystalline peaks formed shows the increase in mechanical strength of the modified catalyst. The SEM results shows the change in surface texture of the fresh catalyst from smooth and uniform to rough and agglomerated.

5.0 Acknowledgement

The research team acknowledge and appreciate the Tertiary Education Trust Fund for providing part funding for this research work

References

Choudary N.V. & Newalkar, B. L. (2011). Use of zeolites in petroleum refining and petro-chemical processes: recent advances. Journal of Porous Materials, 18: 685-692.

Foster A.J., Jae J., Cheng Y. T., Huber G. W., Lobo R.F. Optimizing the aromatic yield and distribution from catalyst fast pyrolysis of biomass over ZSM-5.*Appl. Catal. A* Gen. 2012, 423-424, 154-161.

Hossein J, Halim M, Jelas MD, Rafiee R, Shameli K, Gharibshahi E. Synthesis and characterizarion of zeolite/Fe3O4 nanocomposite by green quick precipitation method. Dig J Nanomater Biostruct 2013; 8(4):1405–13.

Jin, R.; Liu, Y.; Wang, Y.; Cen, W.; Wu, Z.; Wang, H.; Weng, X. The role of cerium in the improved SO2 tolerance for NO reduction with NH3 over Mn-Ce/TiO2 catalyst at low temperature. Appl. Catal. B 2014, 148, 582–588.

Kianfar E. et al., International Journal of Chemical Reactor Engineering, 17(2):1-10 (2019).

Letzsch W., Hart's Fuel, 2010, http://www.hartfuel.com/f.catalyst.html, accessed February 2015; J. Scherzer, Octane Enhancing Zeolitic FCC Catalysts: Scientific and Technical Aspects, Marcel Dekker, New York, 1990, p. 157; P. Niccum and R. Northrup, Proceedings 2006 NPRA meeting, Salt Lake City, paper AM-06-18.

Lopes J.H., Nogueira, F.G.E, Gonçalves, M., Oliveira, L.C. (2015). Modified Zeolite with Transition Metals Cu and Fe for Removal of Methylene Blue from Aqueous Medium: Mass Spectrome-try Study. Bulletin of Chemical Reaction Engineering & Catalysis, 10 (3): 237-248. (doi:10.9767/bcrec.10.3.8624.237-248).

Onoriode PA, Al-Hajri RS, Jibril BY. Catalytic upgrading of heavy oil using NiCo/y-Al2O3 catalyst; effect of initial atmosphere and water-gas shift reaction. Fuel 2018; 235:736–43.

Paasikallio V., Lindfors C., Kuoppala E., Solantausta Y., Oasmaa A., Lehtonen J. Green Chem. 16 (2014) 3549.

Shuwa SM, Al-Hajri RS, Jibril BY, Al-Waheibi YM. Novel deep eutectic solvent-dissolved molybdenum oxide catalyst for the upgrading of heavy crude oil. Ind Eng Chem Res 2015; :3589–601.

36c Universal Trends in Catalytic Methane Activation in Presence of an Oxidizing Agent

Shengguang Wang, Lars Grabow

Methane conversion technologies are an important aspect of both elimination of greenhouse gas emissions and utilization of natural gas and biomass derivatives. However, because of its strong C-H bonds, methane is hard to break down and convert to other higher valued molecules. A series of oxidizing agents have been used to activate methane, and despite methane activation being the common slow step, the reported optimal catalyst and reaction mechanism differ for each oxidizing agent.

Notwithstanding the stark differences in catalytic behavior, we have identified universal trends in catalytic methane activation in the presence of the different oxidizing agents (O_2, H_2O, CO_2, H_2O_2). Our study using density functional theory (DFT) and microkinetic modeling shows the importance of the harmony between oxidizing agent and catalytic metal, and highlights the required balance between the binding strength and surface coverage to achieve the maximum methane activation rate. Our work also successfully predicts the abnormal size dependence of Pd for methane partial oxidation in a study combining DFT calculations with Wulff construction and kinetic Monte Carlo (kMC) simulations. The interplay between different surface facets of a Pd particle has a pronounced effect on the dynamic formation of surface oxygen species, and in turn, methane activation. The diffusion of oxygen atoms from Pd(100) to Pd(111) is crucial for surface oxygen formation on the (111) facet, although the binding energies on these two facets from DFT calculations are nearly identical. Lowering the rate of oxygen diffusion by increasing the width of the (111) and (100) terraces in our model, correctly reproduces the experimentally observed trend of higher methane conversion rates with increasing nanoparticle size. Overall, our results provide fundamental insights into catalytic methane conversion, which can be leveraged for the design of improved catalysts.

36d Exceptional Catalyst Selectivity Prompts Cycle Length Extension at the Petro Rabigh Monoethylene Glycol Plant, Saudi Arabia

Ammar Bakheet

Rabigh Refining & Petrochemical Company (Petro Rabigh) has achieved an impressive 92% catalyst selectivity at the start of the most recent cycle at its 600,000-t/y monoethylene glycol (MEG) production facility in Saudi Arabia. The plant uses the advanced Shell OMEGA process, which is based on the catalytic conversion of both ethylene to ethylene oxide and ethylene oxide to ethylene glycol.

44a Innovative Solution to Select Compatible Process & Material For New & Strange Process streams

Muhammad Farooq Fareed

The metallurgy selection for the compatible process is quite tricky and complex, especially for the new technology and process. The selection would be more difficult once the process has behaved different fluid patterns with nature. There were number of failure happen around the globe only for wrong material choice. It shall not limited for production loss, but affect around billions of dollars assets as well as lost time injury. The same issue faced in one of the affiliate and found the major equipment failure with the huge production loss until permanent solution arrived. The loss not belongs to production but high investment on permanent solution, temporary solution until permanent solution arrival, energy, utility & product loss due to frequent shutdown & startup, threat to worker nearby that equipment and spoil the utilities due to some equipments process failure can release to some utilities. The process is different acidic (two different kind of acids, one is weak and second is strong) nature, the molality of the fluid is beyond the imagination, other supportive element availability can exaggerate the failure phenomena and high temperature and pressure are the contributor for the failure. After the first failure and detail investigation, the above contributor as well as the fluid reveal the behavior of fluid attach. When most of the alloy of steal and other high corrosive resistance metal alloy were failed with such uncertain behavior, then extensive process and material investigation were required to provide the true durable solution required to stop safety threat and production loss. Process and material team extensively worked on it and delivered the new innovative solution compatible with process. This modification was not only limited to material but also had number of process and equipment geometry were modified. The capital cost of the solution was little higher, but saved the major losses & threat, and reduce sustainability figure.

44b New Approached Used to Redesign & Optimize the Flare In Petrochemical Plant

Muhammad Farooq Fareed, Saud Al Jebreen

The United (SABIC affiliate) is design to produce number of petrochemicals, which have different in nature. Every element/compound reactivity has vary to other. Selection of right process and operation with computability of fluid and process conditions. Wrong design & selection will not only disturb the operation but also can be catastrophic. One of the olefin plant is design is produce number of valuable products and rest of heavier product divert to the another SABIC affiliate for further processing. Outage/upset/emergencies of heavier olefins product users plant will force to company to divert this product to the flare. After the few flare events, the operation started to face the some backpressure issues at plant battery limit, later that turn severe. During schedule shutdown, the two flare tips found damaged, especially the last two as well as the some sticky material found in from knock drum till flare sub header assembly. Further investigation and study found the number of reveal about the system. The system have additional injection utilities, we cause such polymerization, which actually not considered during design. The flare system had design with conventional approach. Since the actual design had not considered proper tactics, which ultimately cause downstream equipment failure. By using facts from the investigation, the new approach have developed and used to redesign the complete unit. Which have completely removed the situation and overcome the problem. The approach were used to synchronize two Aspen softwares and interconnected the properties. First time, the reactor & associated utilities used to design the KO vessel and complete flare downstream equipments with synchronization with flare net. With such approach and correction, we have enhanced the complete flare system and life of equipments. The company plant emergency shutdown threat have removed due to high backpressure and keep plant on smooth operation. Those improvement ultimately save the major production, maintenance and replacement cost as well as enhance sustainability & reliability figures.

44c Optimization of Solvent Deasphalting Process to Produce Deasphalted Oil for Partial Upgrading of Heavy Oil with Visbreaking

Youngdoo Kim, Sun-A Choi, Eun-Hee Kwon, Nam Sun Nho, Kwang Ho Kim, Sukhyun Lim, Ki-Duk Kim, Hung Pham

The solvent deasphalting(SDA) process is a partial upgrading process that separates asphaltene to obtain deasphalted oil(DAO) from heavy oil such as bitumen in oil sand. Maximize the yield of DAO from the heavy oil and produce the product that meets the condition of the pipeline transportation, the experimental study of SDA process and configuration with visbreaking process was investigated in this work. This study was conducted in a bench-scale SDA process which consists of feed and solvent supply, extraction column, pitch collection and solvent separation with recovery. The experiment was carried out using effective parameters such as temperatures, pressure, solvent to oil(S/O) ratio and types of solvents. All experimental conditions were adjusted to reduce viscosity and density without consideration of substances such as metals, sulfur and carbon conradson residue in order to satisfy pipeline transport specifications. The result show that removal efficiency of asphaltene increase when using a solvent with a high carbon content. The density and viscosity of DAO was approached to meet pipeline transport specifications. However, the simple combination of the SDA process and visbreaking process still not met the requirement for pipeline transportation. Thus, optimized the operating state of the SDA process with configuration with the vesbreaking process was suggested in this study.

44d CO_2 Dissociation over Ni with Doped Ceria Catalysts

Phillip Mynarski, Saurav Kumar, Ben Jang

The dissociation of CO_2 to CO is a critical step of the chemical looping process that utilizes two major greenhouse gases, CO_2 and methane, with the solar energy. This study focuses on using temperature programmed reduction with various hydrogen/methane mixes followed by a pulse CO_2 dissociation reaction to characterize and evaluate the synthesized catalysts. The pulse reaction activity and stability of catalysts are summarized and compared. There are different Ni loadings of Ni-$CeZrO_2$ used for the study of activity, capacity, and stability for CO_2 dissociation. Characterization of the catalysts includes surface area, oxygen vacancy, and catalyst composition measurements.

44f Predicting the Potential of Asphaltene Deposition at Reservoir Condition in Lower Fars

Adel Elsharkawy, Nora Borashid

Flow assurance plays a key role in effective production management. It is closely related to problems with solid deposits. Solids derived from fluid hydrocarbons such as waxes, asphaltenes, and hydrates. The occurrence of asphaltene deposition is one of the major problems that hinder hydrocarbon production. It can occur anywhere between the reservoir and the surface. Therefore, it is imperative that there be a procedure for minimizing the deposition of asphaltene in the production flow lines.

There are several screening criteria that can be considered to determine the asphaltene deposition. This research will conduct a study on different methodologies or categories to determine the stability of asphaltene: the first category is the stability criteria using indices such as: the colloidal instability index (CII), the colloidal stability index (CSI), the stability index (SI), and the asphaltene stability predicting model (ANJIS). These four methods are used to measure the stability of asphaltene using SARA analysis. The fifth one called modified CII which uses the compositions analysis of reservoir fluids, instead of SARA analysis, to determine the stability of asphaltene. The second category uses SARA Analysis to predict the stability via plots such as: Stankiewics plot (SP), De Boer plot, Sepulveda stability criterion (SCP). The third category is using compositional analysis of reservoir fluids and the equation of state model to predict the asphaltene onset pressure (AOP). In the last method a commercial software was used (PVTSIM) and Soave Redlich Kwong (SRK) equation of state model was used to predict asphaltene precipitations.

In this work, reservoir information, depth, compositional analysis, and SARA analysis from Lower Fars formation is provided by Kuwait Oil Company. These data are used to study the potential of asphaltene deposition at the reservoir level.

The stability criteria using indices and plots shows discrepancy in the results. The only two methods that shows all the datasets are stable are the modified CII and the EOS model.

As far as predicting asphaltene stability is concerned, the EOS method, based on compositional analysis and the thermodynamic model, appears to be the most reliable and trustworthy method. The other methods presented are observational rather than

thermodynamic. Using these criteria is intended as a quick and fast estimate of the stability potential of asphaltene, not for planning or management of asphaltene deposition issues.

44g Impact of Secondary Reactions of Fluid Catalytic Cracking on the Yield of Light Olefins

Shatha Alabbad, Khalid Almajnouni, Isidoro Morales Osorio

Fundamental understanding of the kinetics and mechanisms of essential reactions in catalytic cracking is still not well-understood and has long been recognized as an imperative task in academia and industry. To keep up with the demand of light olefins production while minimizing both the energy intensity and carbon footprint per high-value chemicals produced, as an example, it is necessary to rationally resolve the puzzle behind controlling the excessive formation of paraffins in cracking processes. The presentation will cover our efforts to explore, quantify, and control the occurrence of side reactions, namely hydrogen transfer reactions, in catalytic cracking. A model component, cyclohexane, is used as a probe molecule in a riser simulator to determine the contribution of hydrogen transfer reactions to coke formation and lost olefins production. The reactions are studied over a range of temperatures and residence times. Thus far, surprisingly, initial findings in analyzing results show a paraffins to olefins ratio of more than 3 at 450 °C. The ratio decreases at elevated temperatures reaching, nearly, 1 at 550 °C. Approximately 65 to 40% of formed olefins are re-consumed in hydrogen transfer reactions over the temperature range in which a naphthene donates hydrogen to an olefin and becomes aromatic whereas the olefin is hydrogenated into a paraffin. The data indicates that the contribution of hydrogen transfer reaction is very significant in the conversion of cyclohexane; it is not similar to paraffins cracking. In addition, the investigation results suggest that steaming the catalyst causes an improvement in the production of olefins and in the reduction of hydrogen transfer reactions. During the presentation, I would like to further illustrate the impact of catalyst modification and reaction conditions on the rate of hydrogen transfer reactions.

44h Study on Hydroprocessing of Waste Plastic Pyrolysis Oil for Chemical Recycling

Sukhyun Lim, Eun-Hee Kwon, Hung Pham, Kwang-ho Kim, Ki Duk Kim, Nam Sun Nho

With the growing demand for plastics around the world, interest in plastic waste management is growing. However, the most plastic waste is going to landfills or natural environment. Therefore, the development of technology for plastic recycling is essential, and the technology for chemical recycling is concentrated, especially to solve environmental problems or net zero carbon emission. In this study, for the chemical recycle of waste plastics, we conducted a process technology development to hydroprocess the pyrolysis oil obtained from the waste plastics processing company in south korea and put it into the NCC process, which is the existing petrochemical process. The experiment conducted research on the effects of temperature, which is the main operating variable, in a bench-scale continuous equipment, and the physical properties and composition changes in feedstock and products. As a result, the impurity(N, Cl, O) removal ratio was obtained by up to 90%.

44i Surfactant–Polymer Flooding in Carbonate Using Novel Synthesized Zwitterionic Surfactant and Oilfield Polyacrylamide

Syed Hussain, Muhammad Shahzad Kamal, Afeez Gbadamosi, Shirish Patil, Dhafer Al-Shehri

After waterflooding process, an enhanced oil recovery (EOR) process is applied to increase oilfield recovery performance by improving sweep efficiency and/or pore-scale displacement in the reservoir. Nonetheless, conventional surfactant precipitates in high salinity and high temperature conditions. Herein, a novel synthesized zwitterionic surfactant with sterling salinity and thermal tolerance is commingled with oilfield polyacrylamide for surfactant–polymer flooding process at high-temperature and high-pressure conditions in carbonate core.

Hydroxybetaine zwitterionic surfactant was synthesized for this study. The synthesized surfactant was investigated for its efficiency at lowering interfacial tension (IFT) at the oil-water interface using a spinning drop tensiometer. Besides, the wettability behavior of the surfactant at the rock-fluid interface was investigated via contact angle measurement using a Biolin Scientific drop shape analyzer. Furthermore, the rheology of the surfactant with oilfield polyacrylamide was studied using an Anton Paar rheometer. Finally, an oil displacement test was carried out using high-temperature high-pressure core flooding equipment.

The novel synthesized surfactant was stable in high temperature and high salinity conditions and showed no sign of precipitation. Moreover, the surfactant demonstrated moderate efficiency at lowering the IFT of oil-water to 1.5 mN/m in the presence of seawater salinity and high temperature condition. Furthermore, the surfactant changes the wettability of the rock-fluid interface from oil-wet (165) to aqueous-wetting condition (50). The blend of the zwitterionic surfactant and oilfield polyacrylamide showed good compatibility in seawater salinity and exhibited pseudoplastic behavior at a high shear rate. The slug-wise injection of the surfactant–polymer flood at elevated temperature and pressure conditions yielded higher oil recovery over conventional waterflooding and polymer flooding process. The mechanism of the additional oil recovery recorded is elucidated.

Conventional surfactants precipitate at high salinity and high temperature conditions, thus inhibiting their effectiveness for EOR application. The novelty of this study is the application of novel hydrooxybetaine zwitterionic surfactant with oilfield polyacrylamide for high-salinity and high-temperature carbonate reservoirs.

44j Kinetic Modeling of CO_2-Assisted Oxidative Dehydrogenation of Propane over CrO_x/SiO_2 Catalysts

Iljun Chung, Jinsu Kim, Hyunmin Oh, Hyunjun Jeong, Jisu PARK, Yongju Yun

CO_2-assisted oxidative dehydrogenation of propane (ODPC) has been proposed as an efficient process for production of propylene and CO_2 reduction due to simultaneous reaction between propane and CO_2. Participation of CO_2 as oxidant enhances the equilibrium conversion of propane by involving directly in the ODPC reaction and coupling the dehydrogenation with reverse water-gas shift (RWGS) reaction. Including the RWGS reaction, the ODPC reaction network contains multiple reactions such as direct dehydrogenation of propane, dry reforming of propane, hydrogenolysis, reverse-Boudouard reaction, and etc. Therefore, constructing the kinetic model with these multiple reactions and evaluating of their contribution are required to describe the ODPC reaction. The Mars-van Krevelen (MvK) mechanism, which contains the redox cycle of lattice oxygen, has been proposed to elucidate the mechanism of the ODPC reaction. Thus, it should be considered for constructing the rigorous kinetic model of ODPC reaction. In this work, we report a kinetic model for the ODPC reaction, based on the MvK mechanism with including the multiple reactions. Notably, estimation of kinetic parameters was conducted and the contribution of major reactions were evaluated. With this lumped kinetic model, the optimal operating condition were suggested as a kinetic point of view. For determining the optimal operating condition as a practical point of view, Global Warming Index(GWI) was introduced and calculated to evaluate the net carbon usage for fuel. Hence, the optimal operating condition as a kinetic and practical view was compared.

44k Effect of Phosphorus Modification over Pt/Al_2O_3 catalyst for Non-Oxidative Propane Dehydrogenation

Jungmok Oh, Namgi Jeon, Iljung Chung, Okkyun Seo, Akhil Tayal, Youngju Yun

Catalytic dehydrogenation of propane to propene is important for the production of high-value chemical intermediates such as polypropylene, propylene oxide, and acrylic acid. Pt-based catalysts are considered to be promising catalysts due to their high catalytic activity for C-H bond cleavage. Here, we report the catalytic properties of phosphorus (P) modified Pt/Al_2O_3 for propane dehydrogenation.

To elucidate the promotion effect of P and its role during the reaction, $1Pt\text{-}xP/Al_2O_3$ (x=1~5, wt %) catalysts were prepared by sequential wet impregnation. The physicochemical properties of the catalysts were investigated by CO chemisorption, x-ray diffraction, in-situ x-ray absorption spectroscopy, propane temperature programmed surface reaction spectroscopy, and CO diffuse reflectance infrared fourier transformation spectroscopy.

The addition of phosphorus to Pt catalysts leads to the formation of Pt-P alloy during 600 °C reduction pretreatment. The formation of Pt-P alloy results in the co-presence of metallic Pt and Pt-P alloy in the Pt nanoparticles of the P-modified catalysts. For the propane dehydrogenation at 600 °C, the unmodified $1Pt\text{-}0P/Al_2O_3$ catalyst showed 38% propane conversion and 83% propylene selectivity. However, P-modified $1Pt\text{-}1P/Al_2O_3$ catalyst showed dramatically increased selectivity of 95% at similar conversion, indicating a synergistic effect of metallic Pt and Pt-P alloy species on catalytic performance for PDH. The addition of more P content leads to a further increase in propylene selectivity, but excess Pt-P alloy species lowered the activity of $1Pt\text{-}xP/Al_2O_3$ catalysts. We found that the metallic Pt is the main active species, but Pt-P alloy species promote the selective formation of propylene by suppressing cracking activation.

44l Effect of Size and Polydispersity of Polyacrylamide Solutions Rheology Under Elongational Flow

Diego Soetrisno, Carina Martinez, Mariah J. Gallegos, Vivek Sharma, Jacinta C. Conrad

Polyacrylamide (PAM) is often used for polymer flooding application in enhanced oil recovery within the porous media of an oil reservoir. The porous media creates an extensional flow field that strongly deforms the polymer macromolecules. Extensional rheological phenomenon such as strain hardening may improve oil recovery due to an increase in capillary number. Therefore, it is important to understand how to control the extensional rheology of PAM solutions for optimum oil recovery. Here, we study the extensional flow properties of PAM solutions via the pinch-off dynamics using a dripping-onto-substrate (DoS) protocol. The DoS protocol measures the extensional properties by dispensing a fluid through a nozzle to form a pendant drop and capturing the pinch-off process as the fluid deposited to a sessile drop on a substrate with a high-speed camera. We use polyacrylamides of various molecular weight and polydispersity to investigate the role of macromolecular properties on the extensional response of polymer solutions. We characterize the extensional properties by calculating the extensional relaxation time λ_E and extensional viscosity η_E from the filament-thinning dynamics. With respect to polymer overlap concentration c^* derived from zero-shear viscosity, we observe that the shear rheology at the same c/c^* is almost identical in dilute regime ($c < c^*$). The extensional rheology shows a stronger polymer size dependence, where larger polymer exhibits longer λ_E and a stronger strain hardening response. Concentration-dependence of λ_E follows a power-law exponent that increases with polymer size in dilute regime and transitions to linear in semi-dilute regime ($c > c^*$). For polydisperse polymer, we observe an earlier transition to linear scaling which corresponds to c^* derived from the weight-averaged molecular weight Mw, indicating the stronger contribution of the larger-end of the distribution. We also observe a similar trend when concentrated colloidal suspensions (volume fraction $\Phi \geq 0.40$) are added in the solutions. Our results suggest that we can tune the extensional response of PAM solution from the polymer size and polydispersity, which may elucidate better polymer flooding formulation for optimum oil recovery.

44m Design of a Steam Condensate Waste Processing System and Creative Reuse of the Condensate in the Oil Refinery

Avinashkumar Karre

The condensate produced from the steam users in an oil refinery was considered waste due to concerns over solid depositions in the heat exchangers and potential leaks from the process side. A creative way of recovering condensate and ways to use in different applications within the refinery is illustrated in this poster. Firstly, three levels of condensate were identified based on pressure of the steam. Those are high, medium, and low pressure condensate. High and medium pressure condensate were flashed in two different pressure vessels, and 150 psig and 50 psig steam is recovered. Recovered steam is recycled back to the steam network. The condensate produced from the 150 psig flash drum is fed to the 50 psig flash drum along with the medium pressure condensate from the source. The net recovered low pressure condensate from the 50 psig flash drum is pumped to utility stations, some low duty requirement users in the refinery, and for maintaining warm temperatures around storage tanks. The net condensate after some degree of cooling is further cooled through an air cooler to near ambient temperature, and further pumped into the cooling tower basin as a partial cooling tower make-up. Recycle of condensate generated additional steam for the refinery as a byproduct and also reduced make-up water cost for the cooling tower.

44n Process Design and Mechanistic Modeling of the per- and Poly-Fluoroalkyl Substances (PFAS) Produced from the Thermal Degradation of Fluoropolymers

Patrick B. Dawson, Augustine H. Kim, Joseph D. Taptich, Juan Lucero, Tessy S. Ritchie, Victor A. Jaffett, Bryan Lagasse, Matthew Gettings, Simuck Yuk, Pamela L. Sheehan, Enoch Nagelli

Process Design and Mechanistic Modeling of the Per- and Poly-Fluoroalkyl Substances (PFAS) Produced from the Thermal Degradation of Fluoropolymers

Patrick B. Dawson, Augustine H. Kim, Joseph D. Taptich[1], Juan D. Lucero, Tessy S. Ritchie[1], Victor A. Jaffett[1], Bryan A. Lagasse[1], Matthew L. Gettings[1], Simuck F. Yuk[1], Pamela L Sheehan[2], Enoch A. Nagelli[1]

[1]Department of Chemistry & Life Science, United States Military Academy, West Point, New York 10996

[2]U.S. Army Combat Capabilities Development Command, Explosive Ordnance Disposal, Demilitarization & Experimental Directorate, Army Futures Command, CCDC-AC, Picatinny Arsenal, New Jersey 07806

ABSTRACT

Per- and polyfluoroalkyl substances (PFAS) are primarily composed of fluoropolymers that have a high resistance to breakdown which allows for high versatility in everyday use but causes these substances to be problematic when disposing them. [1] The Environmental Protection Agency (EPA) is pursuing research that is related to PFAS waste streams when disposed through incineration. [2] This technique disposes of chemicals by combustion at high temperatures, upwards of 1000 to 1400 °C. Recently, the EPA expanded its Contaminant Candidate List to include several PFAS subtypes with thousands of distinct chemicals towards identifying additional PFAS that could potentially require future regulation under the Safe Drinking Water Act. [3] In our study, we are looking at the thermal degradation of polytetrafluoroethylene (PTFE) and neat polymers hexafluoropropylene (HFP) and polyvinylidene fluoride (PVDF) through process modeling in Aspen Plus. The thermal degradation process modeled in a Gibbs Reactor in atmospheric combustion conditions with varying oxygen flowrates and temperatures, ranging in temperatures of 250 to 1000 °C. The process feed stream to study the breakdown of PTFE is modeled as its constituent monomers which include HFP, perfluorobutyl ethylene (PFBE), chlorotrifluoroethylene (CTFE), and perfluoropropyl vinyl ethers (PPVEs) reacting with varying air flow rates of stoichiometric oxygen (O_2) and nitrogen (N_2). Thermogravimetric analysis shows the degradation of PTFE initiates around 260 °C while the significant degradation occurs between 550 °C to 600 °C. [4] In addition, Density Functional Theory (DFT) calculations were also used to gain additional mechanistic understandings on the thermal degradation of relevant fluoropolymer structures. Overall, we aim to determine the type and concentration of PFAS products formed during combustion.

KEYWORDS: Per- and polyfluoroalkyl substances (PFAS), Fluoropolymers, Thermal Degradation Analysis, Combustion, Process Design, Aspen Plus, CHEMCAD.

CONTACT: **Dr. Enoch A. Nagelli, Dept of Chemistry and Life Science**, United States Military Academy, West Point, New York 10996. **Email:** enoch.nagelli@westpoint.edu **TEL:** 845-938-3904.

REFERENCES:

1. Kotthoff, M., Müller, J., Jürling, H., Schlummer, M., & Fiedler, D. (2015). Perfluoroalkyl and polyfluoroalkyl substances in consumer products. *Environmental science and pollution research international*, *22*(19), 14546–14559. https://doi.org/10.1007/s11356-015-4202-7

2. S. Environmental Protection Agency. (2020, February). *Per- and Polyfluoroalkyl Substances (PFAS): Incineration to Manage PFAS Waste Streams.* https://www.epa.gov/sites/default/files/2019 09/documents/technical_brief_pfas_incineration_ioaa_approved_final_july_2019.pdf

3. S. Environmental Protection Agency. (2022, October). EPA Issues Final List of Contaminants for Potential Regulatory Consideration in Drinking Water, Significantly Increases PFAS Chemicals for Reviewhttps://www.epa.gov/ccl/contaminant-candidate-list-5-ccl-5

4. Ochi, Keiji. Kawano, Masahide. Matsuda, Muneaki. & Morita, Masatosi. (2008). Thermal Degradation Products of Polytetrafluoroethylene (PTFE) Under Atmospheric Condition. *Department of Environmental Conservation, Ehime University, Tarumi 3-5-7, Matsuyama, Ehime 790-8566, Japan.* https://docslib.org/doc/10173165/thermal-degradation-products-of-polytetrafluoroethylene-ptfe-under-atmospheric-condition-keiji-ochi-masahide-kawano-muneaki-matsuda-masatosi-morita

44o A Comparative Study of Anti-Fouling Characteristics Using Ni-P and Ni-P-Cu Coated Mild Steel

Suparna Bhattacharyya

Heat exchangers are considered to be one the most commonly available industrial devices. Among its various types Shell and Tube Heat Exchanger are the widely popular ones. The reason can be contributed to its easy and very effective mode of application. The structural material for manufacturing heat exchangers are mild steel as its quite easily available and is capable of vast range of usage. However, industrial fouling and microbial attacks can degrade the material drastically. Electroless coating technique is a great way to minimize these problems. It increases the material product value in long run along with the aesthetic refinement. The present study focuses on the comparative analysis of anti-fouling nature using electroless Ni-P and Ni-P-Cu coated mild steel. The surface morphology and phase structure analysis has been carried forward from SEM, EDX and XRD results. The micro-hardness study along with the coating thickness has been analysed and reported. The anti-fouling study has been carried forward using Test rig experiment in lab scale. The results using both Ni-P and Ni-P-Cu has been recorded and compared in the given study. It forecast and optimize the sample that gives the most suitable result. The potential future scope for the current analysis can be attributed to the scaling up of the process into industrial scale heat exchangers. The process aims to benefit the researchers and industrialist of the current field.

44p Experimental Investigation of Magnetic Field Influence on Viscosity of Waxy Crude Oil Emulsion

Xueying Li, Lei Hou

Magnetic treatment is a widely used physical treatment technique in oilfields. The mechanism of magnetic field on waxy crude oil emulsion has drawn much attention. This study aims to explore the relationship of the change of viscosity of waxy crude oil emulsion and magnetic field and mechanism explanation. The static permanent magnet device was constructed. The viscosity of crude oil and emulsion, oil-water interfacial tension and the distribution characters of emulsion droplets before and after magnetic field were experimentally studied. The results showed that when the magnetic treatment temperature was 58 °C (wax precipitation point), the magnetic field intensity was 200 mT, and treatment time was 10 min, the viscosity of emulsion decreased the most and reached 14.5%. Meanwhile, the viscosity of crude oil and oil-water interfacial tension decreased after magnetic treatment. The droplet distribution did not change obviously after applying different magnetic treatment conditions. The magnetic field reduced the viscosity of the emulsion by reducing the viscosity of crude oil and the oil-water interfacial tension. For waxy crude oil, the viscosity decreased significantly when the magnetic treatment temperature was equal to or above the wax precipitation point. The change rates of viscosity of crude oil presented multi-extremal changes, not strictly increasing or decreasing with magnetic field intensity.

56a Upstream Tank Battery Pressure Management using PLC-based MPC

Brandon Biega, Kenneth Craven, Jered Siefert, Daniel Word, Miguel Estrada, Scott Boyden, Alex Winchester, Tyler Smith

A PLC-based multivariable controller is being used to manage central tank battery pressures and minimize well-head backpressure, while honoring gravity-feed flow requirements and process constraints. The controller adapts to varying well line-ups and rate changes as well as downstream pressure swings, significant valve stiction and significant ambient-temperature swings. The controller was designed and implemented using Lin and Associates' Model-less Multivariable Controller, Deltum-XMC. The first controller was commissioned in less than three days with no need for a step-test or process modeling; operating benefits were immediately recognized. A second controller with the same design was installed on another tank battery with different operating constraints and commissioned before the end of the first week. The project team installed controllers on additional tank batteries.

56b Control Valve Response Modeling and Control

Greg McMillan

Control valves are a significant source of nonlinearity and oscillations in most control loops. Not understood are the details and consequences of the dynamic response of control valves and how proportional-integral-derivative (PID) controller features and tuning can mitigate the problems. Techniques are presented for how to simply model the many complexities of the dynamic response to account for the diversity of control valves including the effects of different actuator, positioner, connections, packing, seat-seal, and internal closure member designs. Common mistakes in the valve specification and testing are noted that are leading to disastrous results that are occurring with increasing prevalence driven by a lack of understanding and a mistaken desire to increase valve capacity, reduce valve cost, and minimize leakage. Methods for better valve specification and testing are detailed. Test results and equations are presented to illustrate the consequences of not addressing the issues. Figures show the effect of valve to system pressure drop ratio on the installed characteristic and hence the control valve flow gain's contribution to the open loop process gain. Equations show how the installed flow characteristic and the resolution and lost motion near the closed position determine the actual valve rangeability. Equations are also presented to estimate the limit cycle amplitude and period for a given valve resolution, lost motion, and flow gain and PID controller tuning. Equations to estimate the effect of valve response and PID tuning on the peak and integrated error for load disturbances are also developed. Figures are presented of test results to show the effect of valve response on flow and level processes and what can be done with PID tuning and key features most notably external-reset feedback to reduce oscillations. Valve position control is presented as a better alternative to split range control to increase the precision and rangeability of a valve system with a large valve for coarse adjustment and a small valve for fine adjustment. A brief synopsis of commonly overlooked variable frequency drive (VFD) resolution, dead band, and nonlinearity limitations is also noted to alert users that a VFD may not be the right solution in terms of control and energy savings. The presentation concludes with a list of best practices for valve response modeling and control including the all-important valve specification process.

56c Process, Instrumentation and Controls Analysis Techniques: Leveraging Base Instrumentation and Controls to Achieve Optimal Plant Performance

Michael Taube, Isuru A. Udugama

Much attention and monetary investment is rightly paid to the physical design and maintenance of a process plant's equipment, such as pumps, vessels, heat exchangers, reactors and distillation towers. However, a significant factor affecting a process plant's performance is the design, and especially the on-going maintenance of, field instrumentation and base-layer control structures. This includes the calibration of field sensors & transmitters, final control elements, as well as the regulatory & "advanced regulatory" control structural design. The condition and reliability of these element are increasingly important as greater reliance is placed upon model predictive control, including AI/ML technologies, and real-time optimization applications. While there are various software products in the marketplace which purport to alleviate the need for manual assessment of field instruments and regulatory control performance, their on-going success is mixed, at best. Therefore, human effort to analyze and assess the condition and quality of instrumentation and controls is still required.

This paper describes a variety of data-analysis techniques which have been found useful in identifying & diagnosing field instrumentation issues, such as sticking valves and calibration errors, assessing actual process behavior of distillation columns and determining proper MV-CV Pairing for regulatory controls. The difference between open loop stable (OLS) and Integrating processes and the effect on PID tuning and, thus, plant performance, is also covered.

121a Challenges in Adding Flare Gas Recovery in Petrochemical Plants

Kamal Banjara

Many petrochemical facilities are moving towards achieving a net-zero emissions facility. Data shows that as high as 3.5 percent of world's natural gas is burned in flares, wasting valuable energy resources and contributing to global greenhouse gas emissions. Both oil & gas and the chemical manufacturing industries are showing more interest in doing feasibility studies to evaluate the addition of flare gas recovery systems (FGRS) or flare gas recovery units (FGRU). These facilities believe that recovering the gas would not only reduce greenhouse gas emissions but also provide the facility with cost savings or revenue gain. However, recovering the flare gas in petrochemical facilities does not come without challenges, including inconsistent flow, composition and other process conditions. This presentation addresses some of the challenges that are associated with adding FGRS in a petrochemical facility.

121b A Conceptual Design of Emission-Free Ammonia and Methanol Production

Ying Liu, Yash Khatanhar, Sujing Wang, Qiang Xu

With the global human population increment, the fast growth of food demand presents grand challenges to the world. Rising anthropogenic CO_2 emissions and global temperature have continuously brought great concerns to the world. Ammonia plays a key role as a fertilizer in the agricultural industry, but the commonly used ammonia production method Haber-Bosch represents a huge part of the CO_2 emissions, which escalates climate warming. This paper has developed a novel conceptual design of emission-free ammonia and methanol production complex to produce multiple products simultaneously without CO_2 emissions. It employs natural gas, air, water, and nuclear energy as the major feedstocks and includes four subsystems: (i) an air separation unit; (ii) a water electrolysis plant; (iii) an ammonia production plant, and (iv) a methanol production plant. Through optimally integrating energy/material streams among four subsystems, the complex can produce ammonia, methanol, and pipeline-ready CO_2, simultaneously without carbon emissions. The whole complex is empowered by nuclear energy. The novel conceptual design of the industrial complex has been validated through rigorous simulations. Its economic and environmental benefits have also been provided.

121c Next Generation Process for Bioethylene Production

Brian Boeger

To address growing concerns on the environmental impact of carbon emissions and societal demands for more sustainable production of everyday goods, transitioning to sustainable feed sources is a promising alternative to today's petroleum-based feedstocks. In this presentation, Axens will provide an overview of a two-step technology pathway, licensed by Axens, to convert lignocellulosic biomass waste, an inedible second-generation feed source, into a sustainable bioethylene product that represents a chemically-equivalent replacement for petroleum-based ethylene.

The first step of this process utilizes Axens Futurol® technology by depolymerizing and fermenting naturally-occurring sugars found in a variety of common second-generation biomasses to produce high purity bioethanol. In the Futurol® process, the biomass feedstock undergoes destructurization to produce a consistent feed for the remainder of the process regardless of the original biomass fed to the unit, maximizing feedstock flexibility to accommodate site-specific availability, including seasonality. The critical steps of the Futurol® technology have been demonstrated at semi-commercial scale, and the technology selected for commercialization at one European biorefinery. For the second step, Axens' Atol® technology provides a simple process to dehydrate varying qualities of bioethanol into polymer-grade bioethylene, while minimizing unintended byproduct formation. The Atol® process utilizes water produced during the dehydration reaction as a liquid recycle to manage reactor temperature and achieve bioethylene yields very near maximum stoichiometric targets. The Atol® process achieved its first successful startup in 2022 producing bioethylene from waste-based bioethanol. Easily integrated into this two-step pathway is the opportunity to capture CO_2 and decarbonize process energy demand through the use of electric heaters, resulting in the production of a carbon-negative bio-based ethylene.

Relying on a vast portfolio of proven, de-risked technologies, Axens offers low-carbon processes to meet today's challenges of reducing GHG emissions in the production of sustainable chemicals.

SPRING23
+19TH GCPS
A Joint AIChE and CCPS Meeting

Green, Blue, Grey, or Brown—Pressure Relief Systems Consideration for Methanol in Decarbonization Processes

Justin Phillips, P.E.
ioMosaic Corporation
1900 St. James Place, Suite #700, Houston, TX 77056
phillips.j.tx@iomosaic.com

Prepared for Presentation at
American Institute of Chemical Engineers
2023 Spring Meeting and 19th Global Congress on Process Safety
Houston, TX
March 12-16, 2023

Green, Blue, Grey, or Brown—Pressure Relief Systems Considerations for Methanol in Decarbonization Processes

Justin Phillips, P.E.
ioMosaic Corporation
1900 St. James Place, Suite #700, Houston, TX 77056
phillips.j.tx@iomosaic.com

Keywords: green, blue, grey, brown, methanol, decarbonization, pressure relief systems

Abstract

Market and governmental demand for decarbonization continues to grow, and it is only natural that our industry responds in kind by exploring and advancing the technologies that allow us to decarbonize. One technique in decarbonization is the production of methanol from H_2 feedstocks combined with captured CO_2. The idea of producing methanol on an industrial scale from these sources is not novel, but it is gaining additional popularity among organizations that use methanol as feedstock and that have vested interests in decarbonization. In practice, the technology *is* new to any engineer not previously familiar with it, and while it is composed of the same unit operations with which we are all familiar, it also presents hazards that we may not be familiar with, including those associated hazards we address in pressure relief system design. This paper will explore the special pressure relief system design considerations for different sections of brown, grey, blue, and green methanol production facilities. E.g., electrolysis, synthesis, distillation, and storage.

1 Introduction

Decarbonization is a broad term for reducing or eliminating carbon (nominally carbon dioxide, CO_2) emissions into the environment. The phrases *carbon capture & storage* (*CCS*) and *carbon capture & use* (*CCU*) often accompany decarbonization and are self-explanatory. Many of us recognize the inherent energetic difficulty of decarbonization because global industry is underpinned by carbon-producing processes: energy production, transportation, and practically every single factory or facility. It is *easy* to decompose substances into CO_2 and release heat while doing it, but it is not so easy to retrieve CO_2 and convert it to something more useful in an energy-effective way. However, because there are new incentives to decarbonize, producing methanol from carbon dioxide hydrogenation technology is gaining popularity and industrial deployment.

The terms *green*, *blue*, *grey*, and *brown* are contextual modifiers describing how a process might include: renewable sources (*green*), hydrocarbon fuels by themselves (*grey*) or in conjunction with CCS or CCU (*blue*), or coal and heavy hydrocarbons without CCS or CCU (*brown*). In the context of methanol production, decarbonization can be achieved by capturing CO_2 for use in feedstock, and by storing the carbon within the product methanol molecule. Clearly, these terms also evoke an emotional response: *Green* and *blue* are considered attractive, like verdant trees beneath an open sky; *gray* and *brown* are considered less attractive, like smoke and filth.

But what does this have to do with pressure relief systems? As with any chemical manufacturing process, pressure relief systems must be considered for methanol decarbonization plants, including the unique characteristics of the feedstocks, intermediates, products.

2 A Crash Course in Decarbonization via Methanol

Methanol, CH_3OH, is the simplest of alcohols and an important chemical commodity used as a solvent, fuel, additive, and precursor for manufacturing formaldehyde, plastics, resins, dimethylether, paraffins, and olefins. The NFPA classifies methanol as a Class IB flammable liquid because it has a flash point lower than 73 °F and boiling point greater than 100 °F [1].

Historically, methanol was created by pyrolysis of wood. In the past century, large scale manufacturing has been achieved with syngas production, where carbon monoxide (CO) is reacted with hydrogen gas (H_2) in the presence of a metal catalyst (e.g., copper). The scientific community recognized the ability to produce methanol from carbon dioxide (CO_2) within the past century also, but industry has not favored this approach because the energy intensity requirements (among other factors) are comparatively greater than that of syngas methanol production. Like syngas production, synthesis from CO_2 also involves a metal catalyst, like copper, but with potential mixes of common or rare metals depending on selectivity and desired activity. In other words, CO is easier to manipulate and react than its stable cohort CO_2. However, because there are emerging social, regulatory, and commercial incentives to decarbonize, methanol production from carbon dioxide hydrogenation technology is gaining popularity and industrial deployment.

CO and CO_2 are the products of numerous processes. CO, for example, is commonly created in steam-methane reformation or gasification of coal and biomass. CO_2 is readily produced from the combustion of hydrocarbons and biological aerobic processes.

Hydrogen gas, H_2, is the other feedstock in the synthetic production of methanol. It is likewise sourced from "multi-colored" processes. *Green* hydrogen comes from electrolysis of water, *blue* hydrogen from, e.g., steam-methane reforming that employs CCS or CCU, *turquoise* hydrogen from methane pyrolysis, *grey* hydrogen from steam-methane reforming without CCS or CCU, and *brown* hydrogen from coal or heavy hydrocarbon gasification without CCS or CCU. Obviously, *grey* and *brown* H_2-producing processes do not exemplify decarbonization. Further, if one sources H_2 from electrolysis powered by an upstream non-renewable, how *green* is it? The author supposes that this is debatable.

The optimum decarbonization scheme for methanol production is one that is, at best, *carbon neutral*, meaning that the carbon is kept in a closed cycle and net carbon to the environment is zero. One example is using methanol as fuel (Figure 1). *Green* methanol is often associated with carbon neutrality because, ideally, *green* H_2 is sourced from renewable sources that are free of carbon (e.g., wind-turbine powered electrolysis) and *green* CO_2 is sourced from the air or from renewable sources before carbon entrapment in a closed cycle. However, it is obvious that CO_2

can be sourced from *blue* hydrocarbon-containing processes that utilize CCS or CCU technologies.

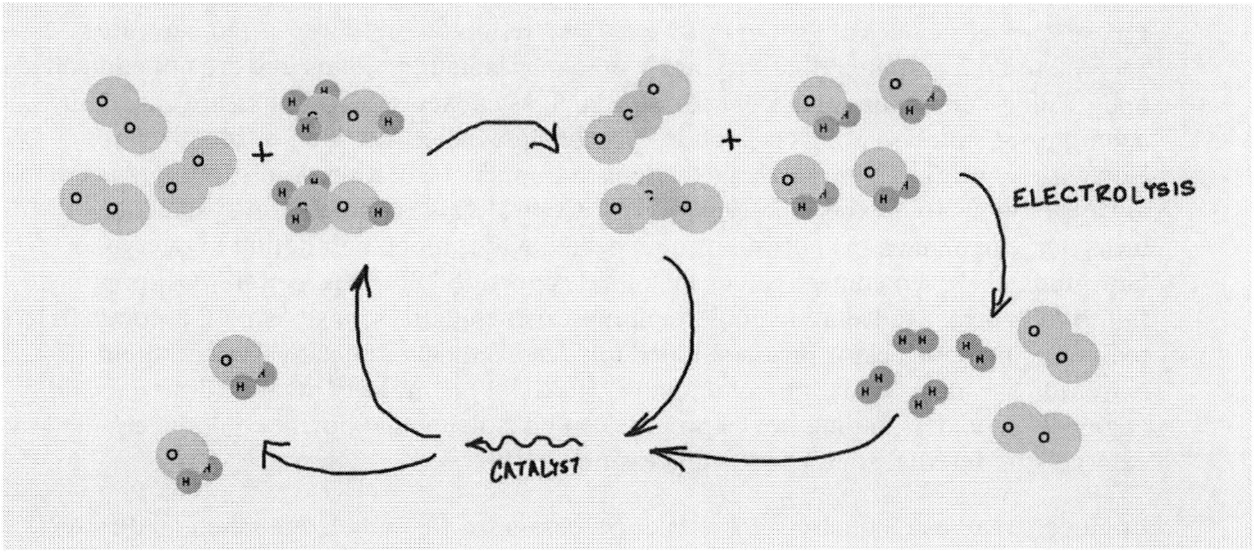

Figure 1: Methanol as Fuel in a Carbon Neutral Cycle (source: ioMosaic Corporation)

3 A Crash Course in Pressure Relief Systems

At a fundamental level, too much heat or too much material flowing into a system (a heat and material *imbalance*) can cause overpressure. Equipment is not infinitely strong, so all contemporary codes of construction, such as ASME VIII [2] for pressure vessels, require protection against overpressure, most often with pressure relief devices. When employed, pressure relief devices are the last layer of defense against a loss of containment from pressure-containing equipment onto people and into the environment. Pressure relief devices may take the form of a reclosing pressure relief valve (PRV), a rupture disk, or even an open vent to atmosphere (where permitted).

For our purposes, a *pressure relief system* is a pressure relief device, its piping, and to a degree its protected equipment (the vessel, tank, rotating equipment, or other equipment that could experience overpressure), and its effluent destination.

Evaluating pressure relief systems is essential for new facility designs and it is very common for *as-built* (existing) facilities. But wait, why would anyone need to evaluate an as-built pressure relief system? Easy question! Because technology improves, because processes change, and because not everything that was originally acceptable remains acceptable.

Engineers evaluate pressure relief systems in three (3) steps:

1) Analyze isolable sections of a process to qualify potential causes of overpressure.

Isolable sections of a process can contain piping, pressure vessels, low pressure tanks, pumps, compressors, heat exchangers, electrolyzers, reactors, and so on. The notion of *isolability* is important because pressure relief system design is intentional and specific. Pressure relief devices are set to relieve at predetermined pressures or temperatures specific to the equipment that they are protecting. Isolation valves that are not controlled open with external administrative controls such as car seals or padlocks are excellent examples of isolation points (which is why they are called isolation or block valves). Likewise, control valves are also excellent examples of isolation points, but this is not always obvious to inexperienced engineers. Control valves and control systems, as it turns out, can behave in spurious or unexpected ways, much befuddling to everyone. Unfortunately, even control valves that are designed to "fail" to a certain position may fail to fail correctly! Rotating equipment may also contain intrinsic isolation points (like a reciprocating compressor piston stopped top dead center) or check valves on their extremities. Piping & instrument diagrams (P&IDs) or similarly detailed drawings aid engineers in understanding how a process can be isolated and how it could be affected by external and internal sources of overpressure.

Engineers may use industry standards or process-specific knowledge when qualifying potential causes of overpressure, but individuals and organizations will find that consistent consideration (i.e., equal treatment of similar systems) is paramount. What is also paramount is the documentation of overpressure qualification, that is, rationalizing why a scenario applies or why it does not apply. These two points (consistent evaluation and documentation of overpressure scenarios) are core to pressure relief system design and are mandated in many parts of industry.

2) Determine the required relief rates for qualified overpressure scenarios.

   After *qualifying* scenarios as applicable, engineers *quantify* them (determine the required relief rate). Industry guidance documents provide common methods to determine required relief rates, which may be based on a simple material balance, an empirical relationship, a natural law, theoretical relationship, or some combination of those things. For example, the required relief rate for a hydrogen gas booster (a type of reciprocating compressor) that has been inadvertently blocked in is equal to the capacity of that booster. For a liquid methanol storage vessel, the rate at which methanol vaporizes due to an external fire is two-part: The heat input from external fire may come from empirical methods underpinned by radiant and convective heat transfer theory. The vaporization rate then comes from the heat input divided by the latent heat of vaporization. On the more "complicated" end of the spectrum, a distillation column that separates methanol from byproducts, waste streams, and recycle streams may experience a reflux failure scenario. For this scenario, engineers create a heat & material balance around the system to determine the imbalance of heat and material compared to normal conditions; excess heat may result in vaporization, and the net product of accumulated heat and material is balanced as a relief load through a pressure relief device.

3) Rate or create the pressure relief systems for the relief requirements. This last step also encompasses effluent handling.

For new facilities, engineers are free to design pressure relief system installations to suit their needs and meet contemporary construction requirements. Frequent iterations and design compromises are common for new facilities—competing interests, constructability, and unforeseen restraints—so it is important for relief designers to remain engaged with piping designers, project engineers, and other stakeholders throughout the design phase because constructing an improper pressure relief design can be costly.

For existing facilities, rating an *as-built* pressure relief system is straightforward: the piping layout, pressure relief device selection, equipment, and process conditions are known (sometimes these parameters are not all known, which complicates matters and increases the effort required to evaluate a pressure relief system). Engineers check the relief capacity of pressure relief devices, ensure that the installation of piping and pressure relief devices is acceptable, and that the design otherwise conforms to good industry practices, such as those known as RAGAGEP. RAGAGEP is a term coined by OSHA that is short for Recognized and Generally Accepted Good Engineering Practice. If the rating of a pressure relief system design is unacceptable, it leads the engineer to consider a more rigorous evaluation ("sharpening the pencil") or to mitigate legitimate design deficiencies in a process design cycle.

Engineers also rate the performance and design of the effluent handling system, which might be a direct discharge to atmosphere or disposal to a complex flare header system. For atmospheric discharge, engineers pay great attention to *safe discharge locations* to avoid harming personnel, equipment, or the environment. The most common materials to atmosphere are "safe," but they are not without their own dangers. E.g., flammable H_2 gas and pressurized steam are discharged to atmosphere frequently. Obviously, locating an H_2 vent away from personnel and equipment mitigates the risk associated with ignited effluent. Locating a steam discharge up and away from personnel mitigates the risk of injuries. For discharge to a closed system, like a thermal oxidizer or flare, engineers must ensure that the disposal system does not create so much backpressure that it reduces the capacity of the pressure relief devices discharging into it. Closed systems provide a safe method of effluent destruction for multiple pressure relief devices—the trick for engineers is to recognize *global scenarios* that place multiple simultaneous demands on the disposal system and cause an increase in backpressure. (Think of this like a traffic jam. In an emergency where you and all your neighbors need to evacuate immediately, undersized roads and small intersections create a resistance to traffic flow, causing you to not move at the required rate.)

4 Tying These Threads Together with RAGAGEP

Many industrial facilities In the United States are subject to federal regulations (laws) that are enforced by the Environmental Protection Agency (EPA) or the Occupational Safety and Health Administration (OSHA), which promulgate consistently safe approaches to process safety, including pressure relief systems. For example, OSHA's PSM Standard [3] is familiar to professionals working in facilities that have more than a threshold quantity (TQ) of highly hazardous chemicals. Both H_2 and methanol are covered by 29 CFR 1910.119(a)(1)(ii) if their quantities are 10,000 lbs or greater.

The author acknowledges that not every facility will be subject to US federal regulations like the PSM Standard. This begs the question: Why mention the PSM Standard? The answer is twofold:

1) Because the PSM Standard is a good model and promotes the concept of RAGAGEP.
2) Because covered facilities that choose to also install pilot-scale *green* or *blue* methanol plants extend PSM coverage to those pilot-scale processes, even if the quantity of H_2 or methanol is less than the TQ.

It is worth noting that H_2 facilities are covered by OSHA's Hydrogen Standard, but it does not fully address pressure relief systems beyond reference to ASME VIII and CGA S-1 [6].

The PSM Standard recognizes that its scope precludes prescriptive requirements for every facility that it covers. Instead, the PSM Standard is performance-based with some simple yet consequential boundary conditions. Namely, those "boundary conditions" are that facilities must document their compliance using RAGAGEP, which are typically consensus standards or best practices. The RAGAGEP indicated in Table 1 are so ubiquitous or relevant that they should not be ignored by engineers performing pressure relief systems work at non-covered methanol production facilities.

Table 1. RAGAGEP Applicable to Methanol Production Facilities

Organization	
Document	**Relevance**
American Petroleum Institute (API)	
API 521: Pressure-Relieving and Depressuring Systems [7]	Standardized overpressure scenario identification and relief rate determination for high-pressure applications; Effluent disposal considerations
API 2000: Venting Atmospheric and Low-Pressure Storage Tanks [8]	Standardized overpressure and vacuum scenario identification and relief rate determination for low-pressure storage tanks; Low-pressure PRD specification and installation
API 520, Part I: Sizing and Selection of Pressure-Relieving Devices [9]	Description of PRD types; Sizing PRDs for single- and two-phase flow
API 520, Part II: Installation of Pressure-Reliving Devices [10]	Installation of PRDs; Pressure drop, backpressure, valve stability, and pipe stress considerations
National Fire Protection Association (NFPA)	

NFPA 2 Hydrogen Technologies Code [11]	Broad application of safety and construction requirements for a range of H_2 facilities; Excludes large scale electrolyzers making ≥100 kg/h H_2
NFPA 30: Flammable and Combustible Liquids	Nearly equivalent to API 2000
American Institute of Chemical Engineers – Design Institute for Emergency Relief Systems and the Center for Chemical Process Safety (AIChE DIERS & CCPS)	
Guidelines for Pressure Relief and Effluent Handling Systems [12]	Rate determination for reactive and nonreactive overpressure scenarios; Sizing pressure-relief devices (PRDs) for single- and two-phase flow; Effluent disposal
National Aeronautics and Space Administration (NASA)	
Safety Standard for Hydrogen and Hydrogen Systems [13]	Contains guidance on vents to atmosphere and to flare systems
Compressed Gas Association (CGA)	
S-1.3: Pressure Relief Device Standards, Part 3, Stationary Storage Containers for Compressed Gases	Description of PRD types; PRD design and installation requirements; Minimum PRD capacity including fire scenario relief rate determination
G-5.5 Hydrogen Vent Systems [14]	Description of H_2-specific vent system from pressure relief devices; Installation requirements
H-3 Standard for Cryogenic Hydrogen Storage [15]	Contains guidance on liquid H_2 pressure relief
Pacific Northwest National Laboratory	
H2 Tools (h2tools.org) [16]	Public website containing myriad H_2 safety guidance including references to relevant standards
Methanol Institute	
Methanol Safe Handling Manual [17]	Contains myriad safety guidance for methanol

5 Special Considerations

New facilities that synthesize methanol from H_2 and CO_2 can be broken into three (3) areas illustrated in Figure 2:

1. Sourcing & Storage

2. Reaction & Separation
3. Product Storage

Figure 2: Three Areas of Methanol Synthesis (source: ioMosaic Corporation)

The author does not intend to include all considerations, but desires to highlight those considerations that may be new to readers that are bridging their experience between traditional hydrocarbon-intensive processes and decarbonization via methanol synthesis.

5.1 Sourcing & Storage

1. There is a potential for solid CO_2 (dry ice) formation in pressure relief systems installed on liquid CO_2 storage, but thankfully, plugging may be unlikely. Leung [18] demonstrated a tendency of "wet" critical CO_2 (contaminated with water) to cause intermittent plugging in discharge piping during relief to atmosphere, but the same did not occur with "dry" CO_2. Experiments by Maher et al [19] were unable to show significant CO_2 solids formation in a commercial-scale pressure relieve valve installation using tortuous 2" discharge piping.
2. Liquid CO_2 relief or pressurized gaseous CO_2 can chill downstream piping systems, so it is important to ensure that downstream piping is free of accumulated water, which can freeze on contact and create a blockage. External freezing of atmospheric moisture (icing on the outside of pipes) is also possible, especially on uninsulated piping.
3. CO_2 is an asphyxiant, so relief to atmosphere must never be to an enclosed area where personnel are present.
4. Liquid H_2 and cold gaseous H_2 pressure relief systems can chill downstream piping systems, instantly freezing any accumulated water and creating a blockage. Furthermore, liquid hydrogen can exit pressure relief devices at temperatures cold enough to freeze atmospheric air. See CGA G-5.5 and H-3 for additional details.
5. H_2 is known for its low ignition energy, flammability, and near-invisible flame. Simply put, H_2 gas tends to catch fire due to static electricity without a passerby even knowing it. NASA suggested incorporating toroidal rings on vent piping exits to minimize the likelihood of static discharge [20]. However, contemporary guidance for smaller,

individual piping in CGA G-5.5 suggests a unique, thrust-balanced vent stack design with mitered cuts. Ignition should always be considered, and although there is general guidance on vent exit location to prevent harm to personnel, a dispersion and radiation analysis is recommended to document the reasonable extent that unignited effluent can travel and the thermal radiation it exhibits after ignition.

6. Despite historical use and any notion to the contrary, using a broom to check for invisible H_2 fires is neither accurate nor especially safe. Specialized adhesives, gas detectors, and thermal detectors are the safest way to check for fugitive H_2 emissions and flames.
7. Electrolyzers, which generate H_2 and O_2 gas from water, come with their own set of special considerations, especially with respect to flammability and explosion hazards. Although often sold as packaged skids that "plug and play" into a facility, responsible owner engineers should take the time to understand the potential causes of overpressure in this technology, including but not limited to external fire on liquid-containing vessels, overproduction of gas, and blocked outlet. Additionally, owner engineers should ensure that pre-packaged pressure relief systems are installed according to RAGAGEP, especially those with atmospheric termination points.

5.2 Reaction & Separation

1. There are several competing methanol synthesis technologies that vary by catalyst selection, reactor cooling schema, and other differentiators. CO_2 hydrogenation to methanol is an exothermic reaction and the potential for a thermal runaway, however slight, must be considered in the pressure relief system design.
2. Consider simplifying pressure relief systems to as few devices as possible for reactor loops by taking advantage of always-open paths, e.g., check if a single pressure relief device can be installed on the reactor effluent separator to protect the separator, reactant coolers, reactor, any recycle compressors, and feed systems.
3. The subsequent separation technology used to separate and purify reactor products may be just as unique as the choice of reactor technology. Facilities may use single, split, or multiple distillation column arrangements, depending on the types of impurities and product streams produced. In any case, there are three important considerations that can be learned from refining and petrochemical industries:
   a. Column imbalance, where the failure of a cooling utility or a blocked outlet can alter the heat and material balance in the system.
   b. Series fractionation, where the failure of a heating utility can cause volatile liquids (or dissolved gas) to leave the bottom and then flash in a lower-pressure, downstream recipient system.
   c. Gas blow-by from the reactor effluent separator, where a loss of liquid level control can cause hot gas or vapor streams from the reactor to enter the distillation system.

5.3 Product Storage

1. Methanol is often stored in low-pressure aboveground storage tanks of API 650 design [21], which may use a fixed roof with an inert N_2 blanket or a floating roof design. Engineers are encouraged to check NFPA 30 and API 2000 for all pressure relief system considerations, which go beyond a generic external fire.

2. Methanol storage pressure relief devices connected to vapor recovery systems must be set low enough to account for frictional losses in the connected piping and high enough to not interfere with normal operation of the tank. Maximal tank design pressures of 2.5 psig (the limit for API 650 designs) provide the greatest margin.
3. Methanol is toxic. In addition to checking for flammable dispersion and thermal radiation of ignited effluent, toxic dispersion analysis from atmospheric discharge points is recommended to ensure that the discharge location is safe. Consider reasonable AEGL exposure times [22].
4. As with the upstream distillation system, consider gas blow-by if storage is interconnected with the separation process.

6 A Glimpse of the Future

AIChE positioned itself years ago to adapt to sustainability and energy transition initiatives across the globe; its Center for Energy Initiatives [23] and Center for Hydrogen Safety [24] are two technical groups that exemplify AIChE's preparedness. Although these groups have their own distinct visions, and purpose, they overlap in the areas of decarbonization and green energy transition. For direct and practical applications, the Center for Hydrogen Safety promotes the use of www.H2Tools.org, a useful and free-to-use web resource developed and maintained by The Pacific Northwest National Laboratory with support from the US Department of Energy.

Other established industry groups that are traditionally hydrocarbon-oriented in their interests, such as the American Petroleum Institute, are following suit. API will continue to lobby for its members' interests in hydrocarbon production and processing, which now includes acting on environmental performance [25], be it reduction of fugitive methane emissions, or reduction in flaring. For the world of pressure relief systems, API offers some of the most influential and widely used RAGAGEP available, such as API 521. API 521 was created with refinery processes in mind, but over the years it has been readily applied to all manner of facilities. Based upon recent meetings within the API Subcommittee on Pressure Relieving Systems, it is likely that API 521 will begin addressing pressure relief in energy transition technologies such as decarbonization with CCS & CCU, electrolysis, and more.

7 Conclusion

This paper attempts to begin bridging the gap between decarbonization via methanol synthesis and pressure relief systems design. It is the author's experience-based approximation that there are few organizations and industry professionals that have deep expertise in decarbonization technologies, methanol synthesis in that vein, and pressure relief system design. The author therefor attempts to promote interest in all these areas by including relevant concepts and technologies such as H_2 generation from electrolysis, methanol synthesis from CO_2 hydrogenation, distillation, Process Safety Management, RAGAGEP, specific standards and best practices, and special considerations.

8 References

[1] National Fire Protection Association. NFPA 30: Flammable and Combustible Liquids Code. 2021 edition. Quincy, MA: National Fire Protection Association; 2021.

[2] American Society of Mechanical Engineers. ASME Boiler and Pressure Vessel Code, Section VIII, Division 1. 2021 edition. New York: American Society of Mechanical Engineers; 2021.

[3] Occupational Safety and Health Administration. Process Safety Management of Highly Hazardous Chemicals. 29 CFR 1910.119. Washington, DC: U.S. Government Printing Office; 1992.

[4] Occupational Safety and Health Administration. Standards Interpretations: RAGAGEP in Process Safety Management Enforcement. May 11, 2016. Washington, DC: Occupational Safety and Health Administration. Available from: https://www.osha.gov/laws-regs/standardinterpretations/2016-05-11. Accessed March 5, 2023.

[5] Occupational Safety and Health Administration. Hydrogen. 29 CFR 1910.103. Washington, DC: U.S. Government Printing Office; 2019.

[6] Compressed Gas Association. Standard for the Marking of Compressed Gas Containers. 9th ed. CGA S-1.3. Washington, DC: Compressed Gas Association; 2020.

[7] American Petroleum Institute. Pressure-Relieving and Depressuring Systems. 7th ed. API Standard 521. Washington, DC: American Petroleum Institute; 2020.

[8] American Petroleum Institute. Venting Atmospheric and Low-Pressure Storage Tanks. 7th ed. API Standard 2000. Washington, DC: American Petroleum Institute; 2014.

[9] American Petroleum Institute. Sizing and Selection of Pressure-Relieving Devices. 10th ed. API Standard 520, Part I. Washington, DC: American Petroleum Institute; 2020.

[10] American Petroleum Institute. Installation of Pressure-Relieving Devices. 7th ed. API Standard 520, Part II. Washington, DC: American Petroleum Institute; 2020.

[11] National Fire Protection Association. NFPA 2: Hydrogen Technologies Code. Quincy, MA: National Fire Protection Association; 2022.

[12] Center for Chemical Process Safety. Guidelines for Pressure Relief and Effluent Handling Systems. 2nd ed. New York, NY: John Wiley & Sons, Inc.; 2017.

[13] National Aeronautics and Space Administration. Safety Standard for Hydrogen and Hydrogen Systems. NSS-1740.16. Washington, DC: National Aeronautics and Space Administration; 1997.

[14] Compressed Gas Association. Hydrogen Vent Systems, CGA G-5.5. 4th ed. Arlington, VA: Compressed Gas Association; 2021.

[15] Compressed Gas Association. Standard for Cryogenic Hydrogen Storage, CGA H-3. 3rd ed. Arlington, VA: Compressed Gas Association; 2019.

[16] National Renewable Energy Laboratory (NREL), U.S. Department of Energy Hydrogen and Fuel Cells Program. H2Tools. Available at: https://h2tools.org/. Accessed March 5, 2023.

[17] Methanol Institute. Methanol Safe Handling Manual. 5th edition. Washington, DC: Methanol Institute; 2020.

[18] Leung JC. Critical CO_2 discharge and visualization. Presented at: DIES Users Group Meeting; September 30, 1992; Princeton, New Jersey.

[19] Maher DW, Valencia, JA, Denton RD, Bevilacqua TJ. Successful demonstration of relieving CO_2-solid-forming streams through a pressure relief system. Mary Kay O'Connor Process Safety Center 18th Annual International Symposium; October 27 – 29, 2015; College Station, TX. College Station, TX: Mary Kay O'Connor Process Safety Center; 2015.

[20] National Aeronautics and Space Administration. Toroidal ring prevents gas ignition at vent stack outlet. NASA Tech Brief 67-10098. Washington, DC: National Aeronautics and Space Administration; 1967.

[21] American Petroleum Institute. Welded Steel Tanks for Oil Storage. API Standard 650. 13th ed. Washington, DC: American Petroleum Institute; 2021.

[22] Environmental Protection Agency. Methanol Results in the AEGL Program. Washington, DC: US Environmental Protection Agency. Available from: https://www.epa.gov/aegl/methanol-results-aegl-program. Accessed March 5, 2023.

[23] American Institute for Chemical Engineers. Center for Energy Initiatives. Available from: https://www.aiche.org/cei. Accessed March 5, 2023.

[24] American Institute for Chemical Engineers. Center for Hydrogen Safety. Available from: https://www.aiche.org/chs. Accessed March 5, 2023.

[25] American Petroleum Institute. The Environmental Partnership. Available from: https://theenvironmentalpartnership.org. Accessed March 5, 2023.

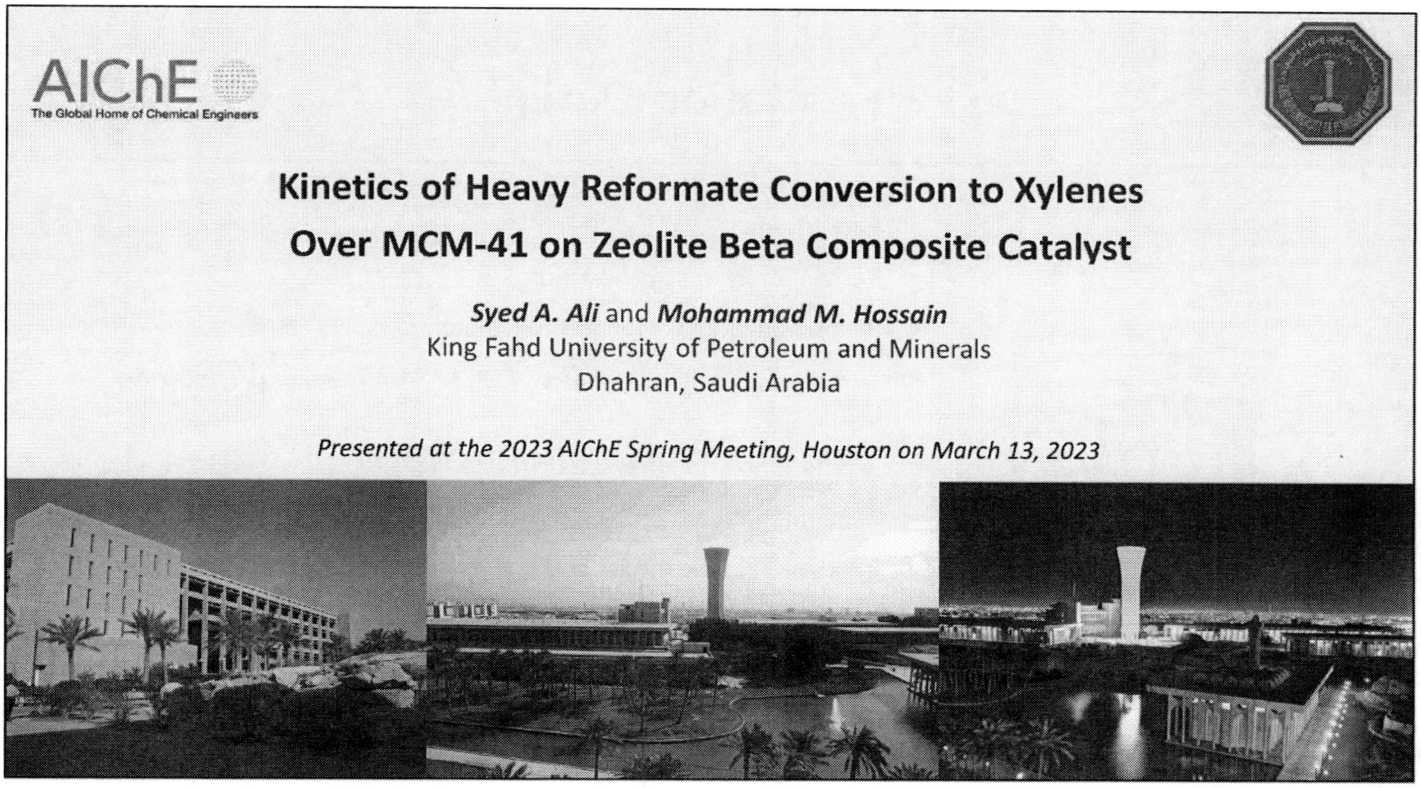

Thank you, Mr. Chairman, for the kind introduction. I feel honored to speak at the 2023 AIChE Spring Meeting here at Houston. Thanks to the organizing committee and Session Chairs for providing me this privilege.

The title of my presentation is" Kinetics of Heavy Reformate Conversion to Xylenes Over MCM-41 on Zeolite Beta Composite Catalyst. My friend and co-author Prof. Mohammad Mozahar Hossain is present in the audience.

King Fahd University for Petroleum and Minerals is located at Dhahran in eastern Saudi Arabia. It is ranked at 160 by QS World University Ranking in 2023. Here are some photographs of our campus.

Presentation Outline

- Background
- Routes for xylenes production
- Reactions/catalysts for the conversion of heavy reformate to xylenes
- Synthesis and characteristics of MCM-41 on zeolite β composite catalysts
- Experimental data for kinetic modeling
- Reaction network and kinetic modeling
- Improved kinetic model and results
- Concluding remarks
- Questions or comments

-2-

Today's presentation will begin with some background information on the subject of the study. Different routes for the production of xylenes will be presented. The chemical reactions that occur during the conversion of heavy reformate to xylenes will be discussed.

Procedure adopted for the synthesis of MCM-41 on zeolite beta composite catalysts as well as their characteristics and performance will be presented.

Methods used for generating he experimental data for kinetic modeling and product composition will be shown.

Reaction network and kinetic modeling, the main part of the study, will be presented along with the steps taken to improve the kinetic model.

I will conclude with some remarks and will be glad to respond to your questions or comments.

Background

Drivers:

- BTX are precursors for many petrochemicals
- Utilization of excess heavy reformate to produce value-added xylenes
- Development of suitable catalyst is the key to enhance the production of xylenes

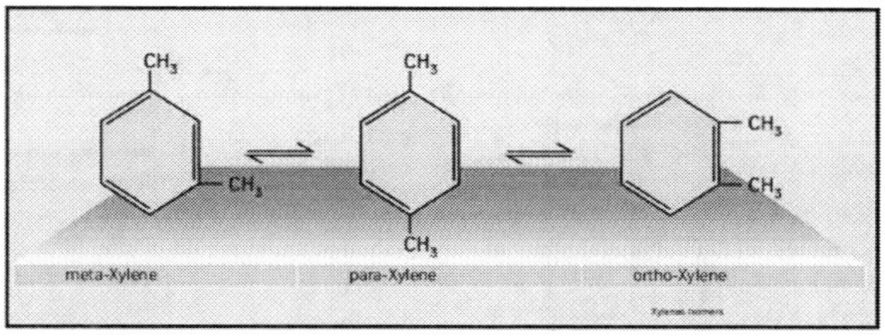

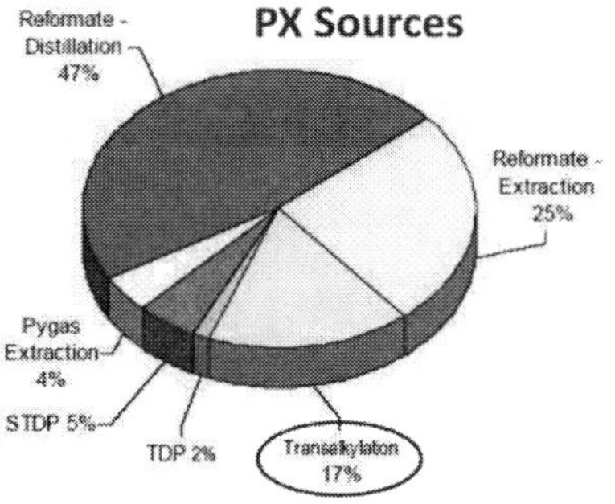

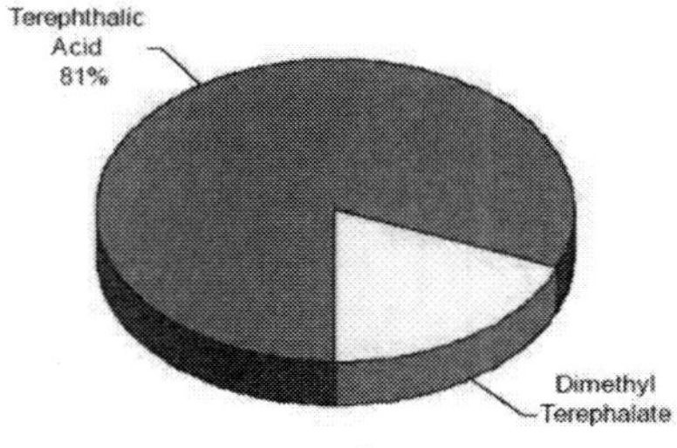

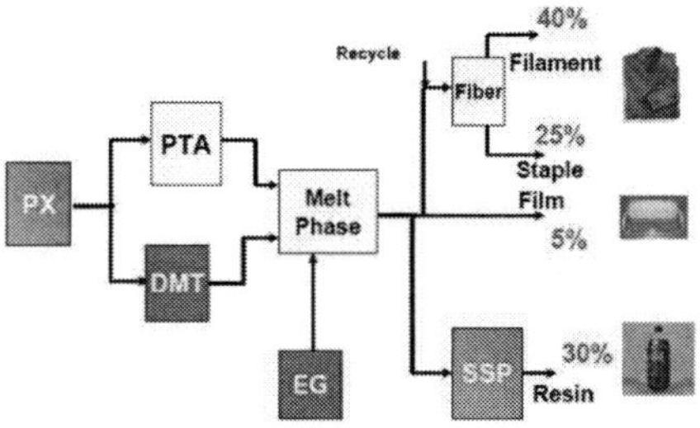

PX End-uses

-3-

As you all know, BTX are the precursors for producing many petrochemicals and their demand (especially the demand for xylenes) is increasing.

At the same time the regulations around the world to reduce the aromatic content in gasoline has resulted in and excess of aromatic-rich heavy reformate.

One of the alternate routes to utilize the heavy reformate is to produce value-added xylenes. Development of suitable catalyst is the key to enhance the xylene production.

Among the isomers of xylene, the demand for para-xylene is exceptionally high due to its use in the manufacture of fibers for textiles, films and resins.

Routes for Enhancing Xylene Production

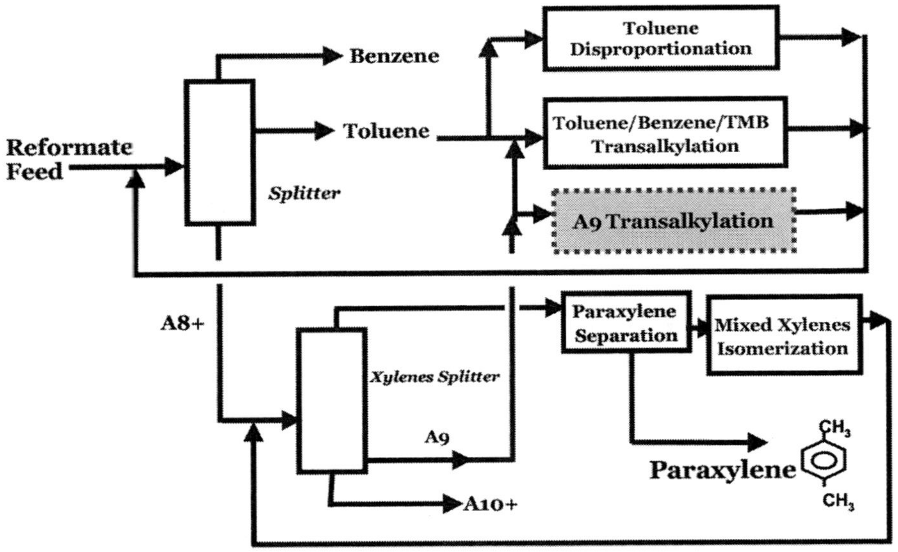

Route	Transalkylation of C9+
Advantages	High xylenes yield Low benzene yield Longer catalyst cycle length Utilization of C10, C11 aromatics
Disadvantages	No toluene utilization Higher hydrogen consumption

-4-

Xylenes are produced by a variety of processes such as toluene disproportionation, transalkylation of toluene or benzene with trimethylbenzenes, or by transalkylation of C9+ aromatics.

Our study is related to transalkylation of C9+ aromatics (without the addition of toluene) which has the advantage of producing higher yields of xylene and low benzene yield. Another advantage is that it utilizes the heavier part of the reformate.

Major Reactions During the Conversion of Heavy Reformate

Heavy Reformate Composition (wt.%)	
o-xylene	0.64
i-Propylbenzene	1.17
n-Propylbenzene	2.80
1-Methyl-2-ethylbenzene	5.18
1-Methyl-3-ethylbenzene	15.01
1-Methyl-4-ethylbenzene	8.60
1,2,3-Trimethylbenzene	6.13
1,2,4-Trimethylbenzene	46.28
1,3,5-Trimethylbenzene	10.07
1,2,3,4-Tetramethylbenzene	0.21
1,2,3,5-Tetramethylbenzene	0.32
1,2,4,5-Tetramethylbenzene	0.35
1,2-Diethylbenzene	0.99
1,3-Diethylbenzene	0.81
Other C10+ aromatics	1.44
Grouped Amounts (wt.%)	
Xylenes	0.64
Methyl Ethyl Benzenes (MEBs)	**28.79**
Tri Methyl Benzenes (TMBs)	**62.48**
C10+ Aromatics	4.12

Reaction	Reactants	Products
Hydrodealkylation	MEB	**Toluene + Ethylene**
	Propylbenzene	Benzene + Propylene
Transalkylation	**Toluene + TMB**	**Xylene + Xylene**
	Toluene + MEB	Xylene + Ethylbenzene
	Toluene + TMB	Benzene + TeMB
Disproportionation	TMB + TMB	Xylene + TeMB
	Toluene + Toluene	Xylene + Benzene
	MEB + MEB	Xylene + Xylene + Ethylene
	MEB + MEB	A_{10} + Ethylbenzene
Isomerization	Xylenes	*o*-xylene, *m*-xylene, *p*-xylene
	TMBs	123-TMB, 124 TMB, 135TMB
	MEBs	1M2E benzene, 1M3E benzene, 1M4Ebenzene
Paring	TeMB	Toluene

Ali et al., Applied Catalysis A, 393 (2010) 96

-5-

Heavy reformate contains a mixture of aromatics – mostly in the C9-C10 range. Trimethylbenzenes (TMBs) are the largest group of compounds, followed by Methyl ethyl benzenes (MEBs).

In the heavy reformate used in this study, TMBs were above 62 wt% while MEBs were about 29 wt.%.

A number of parallel and series reactions occur during the conversion of heavy reformate over zeolite catalysts. Major ones are listed here. Important ones are the dealkylation of MEBs, which forms toluene needed for the main transalkylation reaction produces xylenes.

Thermodynamic Equilibrium Composition

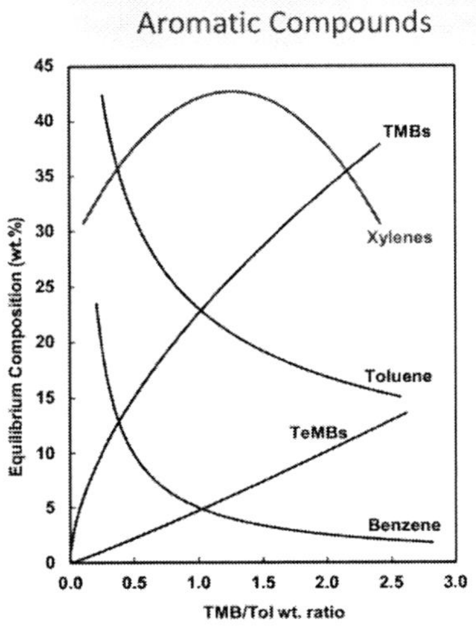

Krejci et al., Appl. Catal. A 377 (2010) 99

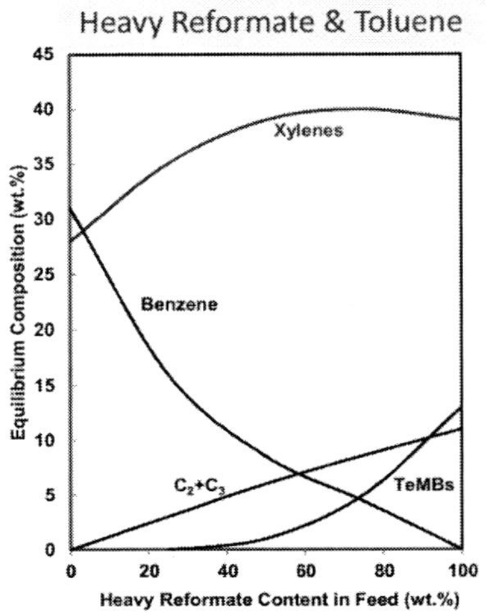

Serra et al., J. Catal. 232 (2005) 342

-6-

The conversion of aromatics is thermodynamically limited. Hence, complete conversion of TMBs is not possible.

When heavy reformate is processed without the addition of toluene, the maximum xylene content in the product can be about 40 wt.% with very small amount of benzene. TeMBs will be less than 10 wt.%. Rest of the product will be unreacted TMBs, MEBs and toluene which can be recycled.

Catalyst Design

- Zeolites with adequate pore size for large molecules (e.g. modernite, β, Y)
- Balance between pore-size, acidity, and metal activity plays a major role
- Ni/mordenite catalysts are used in *Tatoray* and *Transplus* commercial processes
- Problems include faster deactivation, lower conversion of ethyl/propyl- aromatics
- Zeolite β provides better topology than mordenite
- ***Hierarchical composite zeolites*** with secondary porosity in the mesoporous range (2-50 nm) facilitate access of larger reactant molecules to catalyst active sites while maintaining the acidity and crystallinity

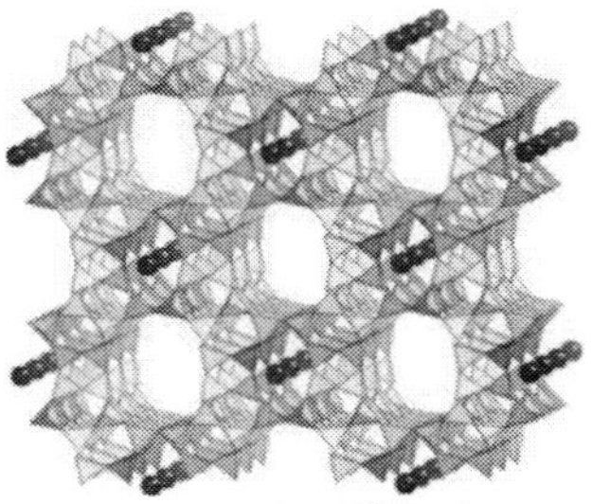

Mordenite
(1D, 12 x 8 MR)

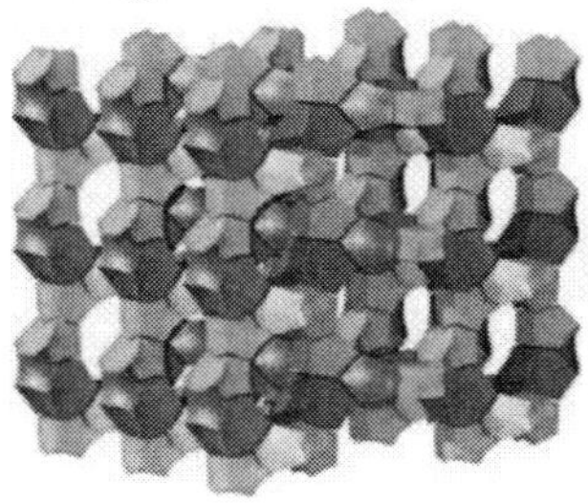

Zeolite β
(3D, 12 x 12 MR)

-7-

Zeolites with adequate pore size are used for the conversion of large molecules of C9+ aromatic. Typically, modernite, zeolite β, or Y-zeolite is used.

A proper balance between pore-size distribution, acidity, and metal activity plays a major role in the conversion of heavy reformate to xylenes,

Commercially, Ni/mordenite catalysts are used. However, they are faced with problems such as faster deactivation and lower conversion of ethyl/propyl- aromatics.

Studies have shown that Zeolite β provides better topology than mordenite because of its large pore volume and three-dimensional structure.

Hierarchical composite zeolites with secondary porosity in the mesoporous range (2-50 nm) facilitate access of larger reactant molecules to catalyst active sites while maintaining the acidity and crystallinity

Synthesis of Metal-Containing Composite Zeolites

- Novel desilication approach involving alkali treatment of zeolite Beta (SiO_2/Al_2O_3 ratio = 38) in the presence of quaternary ammonium surfactant.
- Systematic optimization of the hierarchical pore generation by varying the strength of the alkaline solution during the disintegration of zeolite Beta.
- The strength of NaOH solution was 0.45 M, 0.20 M and 0.10 M for the synthesis of composite zeolites (denoted as **BM45**, **BM20**, and **BM10**, respectively.
- The composites were extruded with Al_2O_3 binder and loaded with **4.0 wt% Mo** by incipient wet impregnation method.
- Catalysts based on **BM45**, **BM20** and **BM10** composite zeolites are denoted as **Mo/BM45**, **Mo/BM20** and **Mo/BM10**, respectively.

Property	Beta (38)	Mo/ BM45	Mo/ BM20	Mo/ BM10
Surface Area				
BET Surface Area (m^2/g)	567	708	714	419
t-Plot Micropore Area (m^2/g)	393	123	325	236
t-Plot External Surface Area (m^2/g)	174	584	389	183
Pore Volume				
Total Pore Volume (cm^3/g)	0.242	0.816	0.598	0.278
t-Plot Micropore Volume (cm^3/g)	0.182	0.052	0.167	0.122
Mesopore Volume (cm^3/g)	0.060	0.764	0.431	0.156
Average Pore Diameter (nm)	2.67	4.61	3.35	2.65
NH_3 TPD Results				
Total Acidity (NH_3 μmol/g)	768	439	668	569
Area of Peak I (100-300°C)	531	341	463	388
Area of Peak II (300-550°C)	237	98	205	181

* Ali et al., J. Ind. Eng. Chem., 94 (2021) 189

-8-

A novel desilication approach was adopted in the presented study to synthesize the composite catalysts. It involves alkali treatment of zeolite Beta in the presence of a quaternary ammonium surfactant.

Systematic optimization of the hierarchical pore generation was carried out by varying the strength of the alkaline solution from 0.45 M, 0.20 M and 0.10 M for the synthesis of composite zeolites (denoted as **BM45**, **BM20**, and **BM10**, respectively.

The composites were extruded with Al_2O_3 binder and loaded with **4.0 wt% Mo** by incipient wet impregnation method.

Catalysts based on **BM45**, **BM20** and **BM10** composite zeolites are denoted as **Mo/BM45**, **Mo/BM20** and **Mo/BM10,** respectively.

The composite catalysts showed higher surface are pore volume – especially mesoporous volume. The NH3-TPD results show reduction in acidity due to desilication.

Performance of Metal-Containing Composite Zeolites

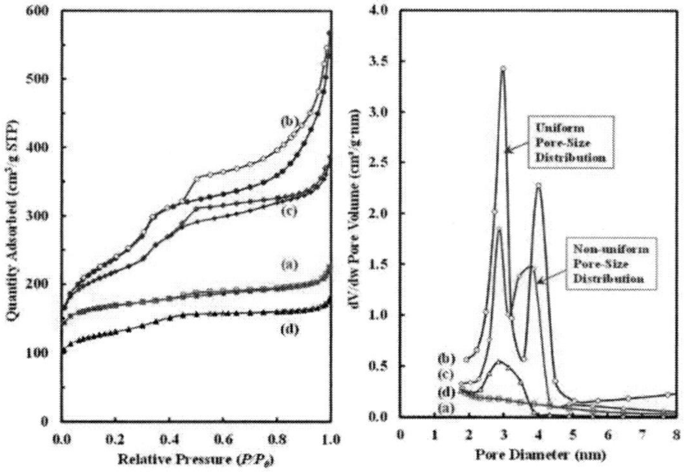

Nitrogen adsorption-desorption isotherms and pore size distribution: (a) **Beta (38)**, (b) **Mo/BM45**; (c) **Mo/BM20**; and (d) **Mo/BM10**.

Catalytic Performance	**Beta (38)**	**Mo/ BM45**	**Mo/ BM20**	**Mo/ BM10**
Conversion (wt.%)				
MEBs	60.40	73.57	85.03	78.33
TMBs	37.80	48.83	51.34	49.97
Yield (wt.%)				
Xylene Yield	20.47	29.36	33.05	29.49
*Xylene Yield Relative to **Beta (38)***	1.00	1.43	1.61	1.44
Selectivity				
Xylene Selectivity	1.08	1.53	1.82	1.40
*Xylene Selectivity Relative to **Beta (38)***	1.00	1.41	1.68	1.29
Ratios				
p-/o- xylene	1.07	0.88	0.88	0.89
p-/m- xylene	0.46	0.39	0.35	0.37

400 °C, H_2: 2 MPa, LHSV: 2.4 h^{-1}, H_2:HC ratio: 4.

Catalyst Mo/BM20 is selected for kinetic modeling

-9-

Bimodal pore-size distribution was found in the composite catalysts with distinct difference from the parent zeolite beta characteristic of well-ordered mesporous MCM-41.

When the catalytic performance of the parent zeolite beta is compared to the composite catalysts, a significant improvement in xylene yield and selectivity is noticed.

The maximum improvement was with Mo/BM20 catalyst with a 60% increase xylene yield and 80% increase in xylene selectivity.

Hence, Mo/BM20 catalyst is selected for the kinetic study.

Experimental Data for Kinetic Study

- Experiments were conducted in **CREC Riser Simulator** which is fluidized batch reactor
- It facilitates **very short reaction times** (5-20 seconds) and accurate control of reaction conditions.
- Internal and external **mass transfer restrictions were eliminated** due to very small size of catalyst particles (15-110 μm) and intense mixing, respectively.
- Experiments were conducted at 300, 350 and 400°C at a catalyst-to-reactant ratio of five for 5-20 seconds.

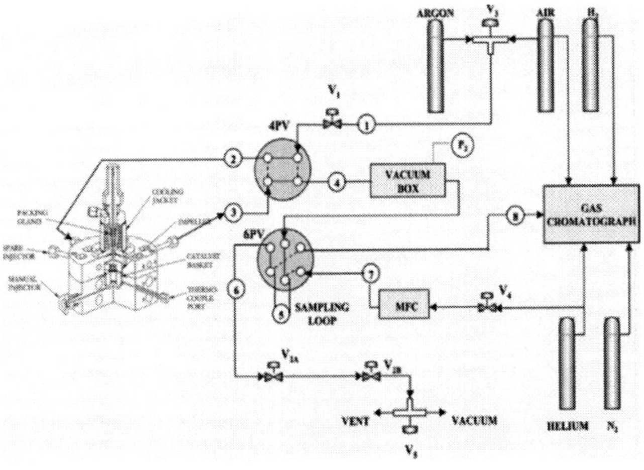

-10-

Experiments were conducted in **CREC Riser Simulator** which is a fluidized batch reactor. It facilitates **very short reaction times** (5-20 seconds) and accurate control of reaction conditions.

Internal and external **mass transfer restrictions were eliminated** due to very small size of catalyst particles (15-110 μm) and intense mixing, respectively. Hence this reactor system is quite suitable for kinetic studies.

Experiments were conducted at 300, 350 and 400°C for 5-20 seconds.

Experimental Data for Kinetic Study

Reaction Temperature (°C)	Reaction time (s)	Feed/Product Composition (wt.%)					Conversion (wt.%)		
		MEBs	TMBs	Toluene	Xylenes	TeMBs	MEBs	TMBs	C9
Heavy Reformate Feed		28.79	62.48	0.00	0.64	0.65	-	-	-
300	5.0	22.58	59.38	2.57	11.11	4.35	27.34	11.83	16.73
	7.5	21.07	57.71	3.35	12.80	5.06	32.20	14.32	19.96
	10.0	19.54	56.01	4.14	14.52	5.79	37.13	16.83	23.24
	12.5	18.63	54.91	4.57	15.60	6.29	40.05	18.47	25.28
	15.0	17.72	53.80	5.01	16.69	6.78	42.99	20.12	27.34
	17.5	16.66	52.52	5.54	18.01	7.27	46.38	22.01	29.71
	20.0	15.63	51.35	6.08	19.17	7.77	49.72	23.76	31.95
350	5.0	20.58	57.53	4.40	13.54	3.95	33.77	14.58	20.64
	7.5	18.82	55.54	5.43	15.60	4.61	39.43	17.54	24.45
	10.0	17.03	53.51	6.49	17.69	5.28	45.21	20.54	28.33
	12.5	15.68	51.98	7.28	19.35	5.71	49.55	22.82	31.26
	15.0	14.32	50.43	8.08	21.03	6.14	53.93	25.12	34.22
	17.5	13.44	49.35	8.54	22.17	6.51	56.76	26.73	36.21
	20.0	12.51	48.09	9.51	23.03	6.86	59.75	28.60	38.43
400	5.0	19.75	56.34	6.08	14.67	3.15	36.44	16.34	22.69
	7.5	16.67	54.14	7.53	17.72	3.94	46.38	19.61	28.07
	10.0	14.34	51.83	9.02	20.07	4.74	53.85	23.04	32.77
	12.5	12.53	50.68	10.10	21.85	4.83	59.67	24.75	35.77
	15.0	11.23	49.00	11.10	23.78	4.88	63.87	27.24	38.80
	17.5	9.52	47.73	12.01	25.68	5.06	69.38	29.13	41.84
	20.0	8.28	45.45	13.78	27.36	5.13	73.36	32.52	45.41

- Formation of negligible amount of benzene (<2 wt.%).
- Thermodynamic investigation shows formation of low amount of benzene when the 100% heavy reformate is processed.
- Hence, the reactions that form benzene are not included in the kinetic model.

-11-

Here is the data used for the kinetic modeling obtained from experiments conducted at 300, 350 and 400°C for short contact times of 5-20 seconds.

The amount of benzene formed was < 2wt%. Thermodynamic investigation shows formation of low amount of benzene when the 100% heavy reformate is processed.

Hence, the reactions that form benzene are not included in the kinetic model.

Conversion of MEBs (a), TMBs (b) and yield of xylenes (c)

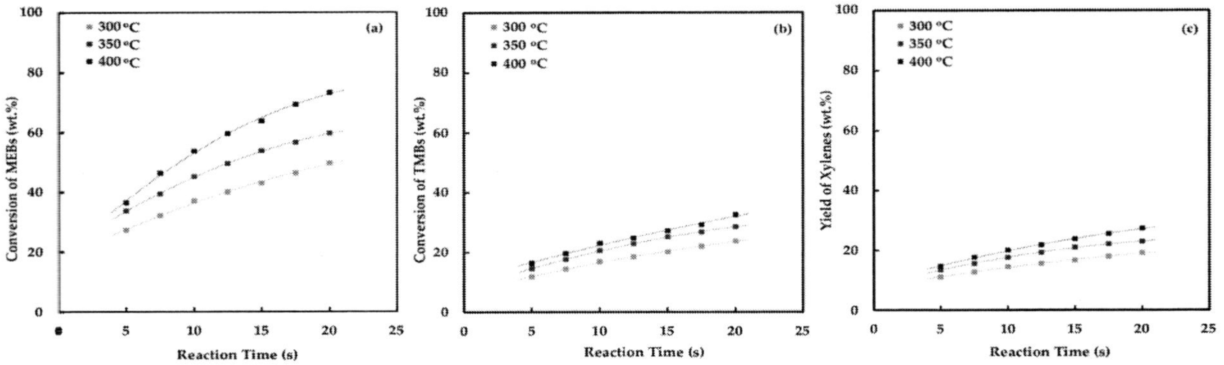

- Conversion of more reactive MEBs was more than two times that of conversion of TMBs
- TMBs conversion (<33 wt.%) was in the kinetic regime (< 52 wt.% is thermodynamic limit)
- Xylene yields increased modestly from 19.1 wt.% at 300 °C to 27.4 wt.% at 400 °C in 20 s.

-12-

Here are some plots to check the consistency of the experimental data. Plot (a) shows the conversion of MEBs versus reaction time and the plot (b) shows the conversion of TMBs. Conversion of MEBs was more than two times that of conversion of TMBs, because MEBs are more reactive and their amount is less compared to TMBs.

TMBs conversion (<33 wt.%) was in the kinetic regime (< 52 wt.% is thermodynamic limit)

Xylene yields increased modestly from 19.1 wt.% at 300 °C to 27.4 wt.% at 400 °C in 20 s.

Selectivity towards xylenes (a), toluene (b) and TeMBs (c)

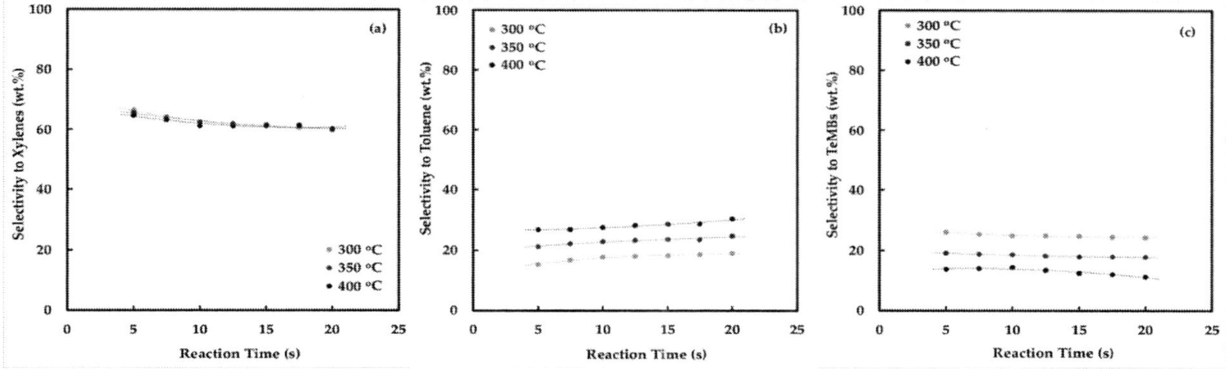

- Selectivity to xylenes is quite high (60-65 wt.%) and remained almost unchanged.
- Toluene is produced by dealkylation of MEBs and consumed in transalkylation reaction.
- TeMBs are produced by disproportionation of TMBs and consumed in paring reaction.
- TeMBs selectivity decreased with temperature because of higher rate of paring reaction.

-13-

These plots show selectivity of different products.

Selectivity to xylenes is quite high (60-65 wt.%) and remained almost unchanged with changes in reaction temperature or time.

Toluene is produced by dealkylation of MEBs and consumed in transalkylation reaction. It increases with temperature and time.

TeMBs are produced by disproportionation of TMBs and consumed in paring reaction.

TeMBs selectivity decreased with temperature because of higher rate of paring reaction.

Formulation of Kinetic Model for Heavy Reformate Conversion

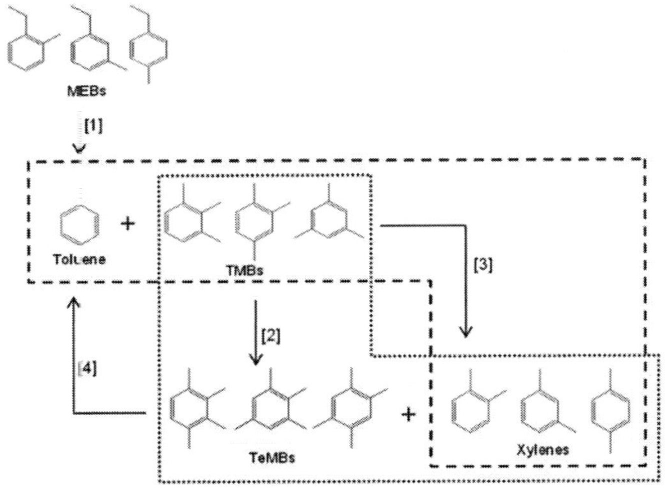

Al-Mubaiyedh et al., Chem. Eng. Res. Des., 90 (2012) 1943

Simplified 4-Reaction Network

[1] Dealkylation of MEBs to form toluene

[2] Disproportionation of TMBs to form TeMBs and xylenes

[3] Transalkylation of toluene and TMBs to form xylenes

[4] Paring reaction of TeMBs to form toluene

-14-

Okay, so now we can proceed with the next part of the study which is kinetic modeling. To begin with, a simplified reaction network with 4 major reactions is used for kinetic modeling.

[1] Dealkylation of MEBs which produce in-situ toluene

[2] Disproportionation of TMBs to form TeMBs and xylenes

[3] Transalkylation of toluene and TMBs to form xylenes –the main reaction.

[4] Paring reaction of TeMBs to form toluene

This reaction network was earlier used by our group while studying the kinetics with other catalysts based on zeolites other than zeolite beta.

Formulation of Kinetic Model for Heavy Reformate Conversion

To further simplify the kinetic model, some assumptions were made, such as :

- Reactions are **elementary**;
- **Single catalyst deactivation** function for all the reactions; and
- **Isothermal** reaction conditions.

The intrinsic reaction rates were correlated with species mole balance with the following assumptions:

- Occurrence of **only catalytic reactions** and thermal conversion is not taken into consideration.
- At any given instant, the **concentrations of chemical species are uniform** due to the high intensity of gas recirculation and a short reaction time.
- The concentration of **coke is uniform** due to the high degree of fluidization at any given time.

-15-

As it is normally done, we made some valid assumptions to further simplify the kinetic model, such as :

- Reactions are **elementary**;
- **Single catalyst deactivation** function for all the reactions; and
- **Isothermal** reaction conditions.

The intrinsic reaction rates were correlated with species mole balance with the following assumptions:

- Occurrence of **only catalytic reactions** and thermal conversion is not taken into consideration.
- At any given instant, the **concentrations of chemical species are uniform** due to the high intensity of gas recirculation and a short reaction time.
- The concentration of **coke is uniform** due to the high degree of fluidization at any given time.

Formulation of Kinetic Model for Heavy Reformate Conversion

Rate of disappearance of MEBs: $\frac{V}{W_c}\frac{dC_{MEBs}}{dt} = -(k_1)C_{MEBs}\varphi$

Rate of disappearance of TMBs: $\frac{V}{W_c}\frac{dC_{TMBs}}{dt} = -\left(2k_2C_{TMBs}^2 - k_3C_{TMBs}C_{TOL}\right)\varphi$

Rate of formation of toluene (TOL): $\frac{V}{W_c}\frac{dC_{TOL}}{dt} = (k_1C_{MEBs+}k_4C_{TeMBs} - k_3C_{TMBs}C_{TOL})\varphi$

Rate of formation of xylenes (XYLs): $\frac{V}{W_c}\frac{dC_{TXYLs}}{dt} = \left(k_2C_{TMBs}^2 + k_3C_{TEBs}C_{TOL}\right)\varphi$

Rate of formation of TeMBs: $\frac{V}{W_c}\frac{dC_{TeMBs}}{dt} = \left(k_2C_{TMBs}^2 - k_4C_{TeMBs}C_{TOL}\right)\varphi$

Catalyst deactivation function φ for the reactant conversion model $\varphi = \exp(-\lambda x)$

Reaction rate constant k_i is a function of temperature (Arrhenius equation): $k_i = k_{i0}\exp\left[-\frac{E_i}{R}\left(\frac{1}{T}-\frac{1}{T_0}\right)\right]$

where
C_i = molar concentration of i, V = reactor volume, W_c = catalyst mass, and t = reaction time, λ = catalyst deactivation constant, and x = conversion of C9 aromatics, k_{i0} = pre-exponential factor, E_j = activation energy of reaction j.

-16-

These are the rate equations for different species based on power law.

A single catalyst deactivation function is included in the model.

Estimation of Kinetic Parameters and Model Evaluation

- Kinetic parameters were evaluated by **least-squares fitting** of experimental data using **MATLAB®**.
- Regression analysis by **Levenberg-Marquardt algorithm** and the differential equations were solved by **Runge-Kutta-Gill method**.
- A total of 105 data points were used for the regression analysis whereas the number of parameters to be estimated was 11 – the **degree of freedom for parameter estimation** (94) was quite high.
- Following principles were also ensured in the estimation of the kinetic parameters:
  - The specific reaction rates and activation energies should **agree with physical-chemical principles**.
  - The **coefficient of determination** (R^2) should be close to unity.
  - The **sum of squares of the residuals** (SSR) should be lower than an acceptable minimum value.
  - The kinetic parameters should display **low confidence interval spans** and exhibit **low-interactions**.

-17-

- Kinetic parameters were evaluated by **least-squares fitting** of experimental data using **MATLAB®**.
- Regression analysis by **Levenberg-Marquardt algorithm**, while the differential equations were solved by **Runge-Kutta-Gill method**.
- A total of 105 data points were used for the regression analysis whereas the number of parameters to be estimated was 11 – the **degree of freedom for parameter estimation** (94) was quite high.
- The following principles were also ensured in the estimation of the kinetic parameters in order to obtain a kinetic model which is valid in terms of the chemical-physical laws and mathematical correctness.

Estimation of Kinetic Parameters

Reaction	MEB Dealkylation		TMB Disproportionation		Toluene-TMB Transalkylation		TeMB Paring		Catalyst Deactivation Function
Kinetic Parameter	k_1 m^3/kgcat.s (x 10^3)	E_1 (kJ/mol)	k_2 m^6/kgcat.s (x 10^3)	E_2 (kJ/mol)	k_3 m^6/kgcat.s (x 10^3)	E_3 (kJ/mol)	k_4 m^3/kgcat.s (x 10^3)	E_4 (kJ/mol)	λ
Value	6.01	21.00	0.029	15.05	1.00	10.11	3.45	96.15	0.79
95% CL	0.817	1.307	0.006	2.87	0.15	1.53	0.321	8.36	0.21

-18-

Here are the estimated kinetic parameters for the four reactions and the catalyst deactivation function. The 95% confidence levels are within the acceptable levels.

Lowest activation energy is for Toluene-TMB transalkylation which is the main xylene producing reaction.

Highest activation energy is for TeMB paring reaction which involve the bulky C10 molecule.

Model Evaluation: Parity Plots and Correlation Coefficients

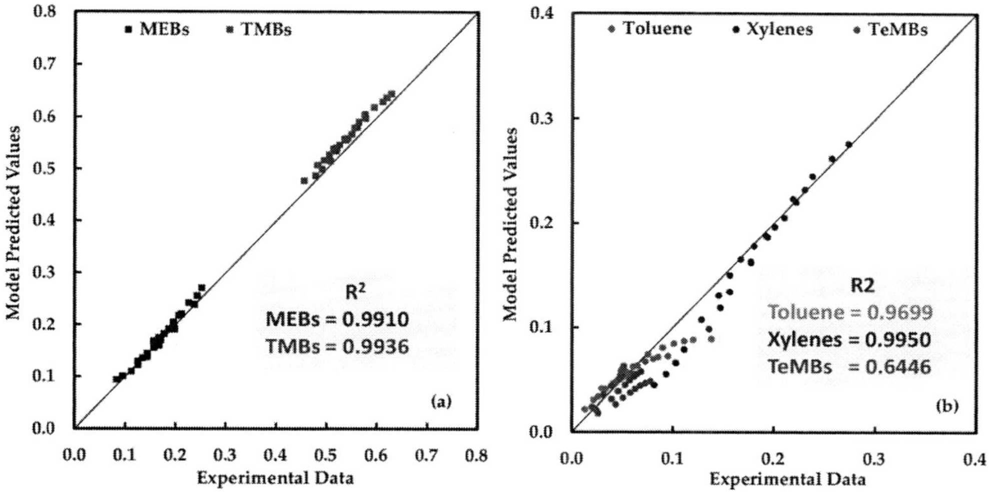

Unsatisfactory prediction of TeMBs and toluene by 4-reaction model

-19-

Let us see how does the kinetic model predicts the composition of products.

Plot (a) shows the parity plot for the unconverted feed components which are MEBs and TMBs. The model seems to predict their composition quite well. Although all predicted values of TMB are slightly above the parity line, R2 values are quite acceptable.

Plot(b) shows the parity plot for the products which are toluene, xylenes, and TeMBs. Clearly, the prediction of TeMBs and toluene is unsatisfactory.

Need to reconsider the kinetic modeling strategy.

The 4-reaction network was based on studies conducted with mordenite. The present study is with composite catalyst based on zeolite beta. Is there any other reaction that is favored by zeolite beta? To answer this question we looked at our previous model compound studies.

Conversion of MEBs to Xylene over Different Zeolites

- An earlier study* investigated the conversion of MEBs over medium-and large-pore zeolites with different topologies and acid-site concentrations.

Catalyst Code	Zeolite Type	Channel Diameter (nm)
Z-150	ZSM-5	0.53 x 0.56 0.51 x 0.55
Z-80		
Z-27		
M-180	Mordenite	0.65 x 0.70 0.26 x 0.57
M-18		
B-25	Zeolite β	0.76 x 0.64 0.55 x 0.55

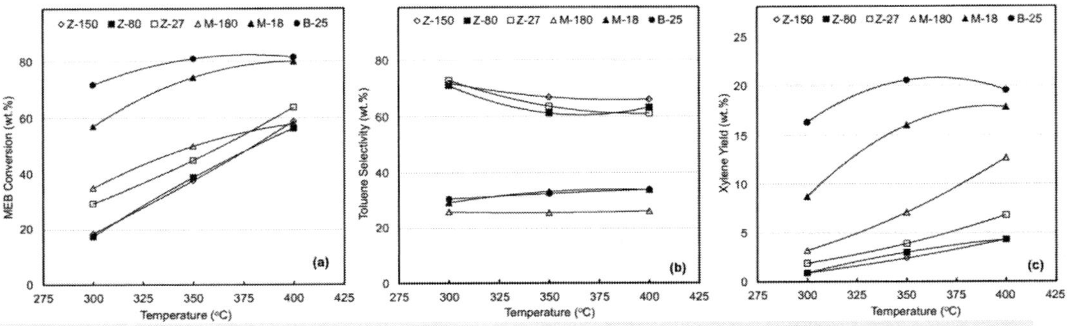

- Results of that study indicated that MEB conversion is significantly influenced by zeolite type and by SiO_2/Al_2O_3 ratio within the same type of zeolite.
- Zeolite beta, a three-dimensional 12 x 12 MR pore zeolite, show a highest MEB conversion (70-80%) with low dealkylation selectivity (30-34%) and high disproportionation and transalkylation selectivity.

* Ali et al., Chem. Eng. Res. Des., 91 (2013) 2601

-20-

In one of our earlier studies, we investigated the conversion of MEBs (as model compound) over six medium-and large-pore zeolites with different topologies and acid-site concentrations.

Results of that study indicated that MEB conversion is significantly influenced by zeolite type and by SiO_2/Al_2O_3 ratio within the same type of zeolite.

Zeolite beta, a three-dimensional 12 x 12 MR pore zeolite, show a highest MEB conversion (70-80%). And shows relatively low dealkylation selectivity (30-34% of toluene) and high disproportionation and transalkylation selectivity (xylene formation).

These results indicate direct conversion of MEBs to xylenes which occurs only over zeolite beta based catalysts due to its large channel diameter and pore volume.

Reformulation of Reaction Network

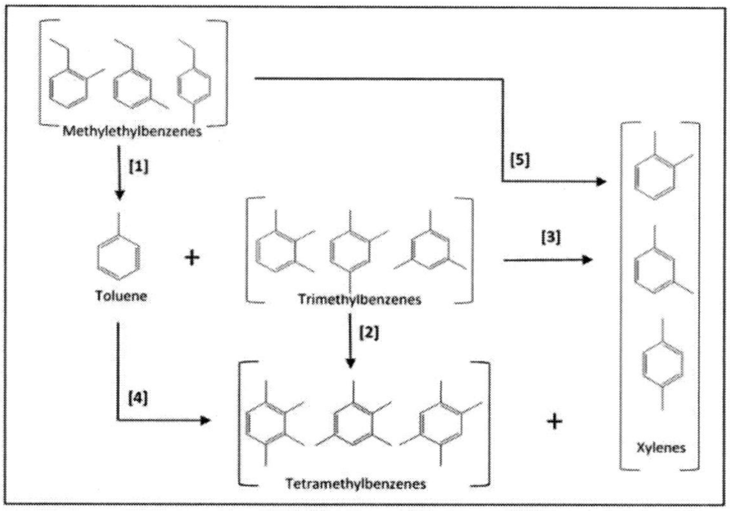

Five (5) Reaction Network

[1] Dealkylation of MEBs to toluene

[2] Disproportionation of TMBs to TeMBs and xylenes

[3] Transalkylation of toluene and TMBs to xylenes

[4] Paring reaction of TeMBs to toluene

[5] Direct conversion of MEBs to xylenes

-21-

The reaction network was reformulated by adding a fifth reaction which is direct conversion of MEBs to xylenes.

Reformulation of Kinetic Equations

Rate of disappearance of MEBs: $\frac{V}{W_c}\frac{dC_{MEBs}}{dt} = -(k_1 + k_5)C_{MEBs}\varphi$

Rate of disappearance of TMBs: $\frac{V}{W_c}\frac{dC_{TMBs}}{dt} = -\left(2k_2C_{TMBs}^2 - k_3C_{TMBs}C_{TOL}\right)\varphi$

Rate of formation of toluene (TOL): $\frac{V}{W_c}\frac{dC_{TOL}}{dt} = (k_1C_{MEBs+}k_4C_{TeMBs} - k_3C_{TMBs}C_{TOL})\varphi$

Rate of formation of xylenes (XYLs): $\frac{V}{W_c}\frac{dC_{TXYLs}}{dt} = \left(k_2C_{TMBs}^2 + k_3C_{TEBs}C_{TOL} + k_5C_{MEBs}\right)\varphi$

Rate of formation of TeMBs: $\frac{V}{W_c}\frac{dC_{TeMBs}}{dt} = \left(k_2C_{TMBs}^2 - k_4C_{TeMBs}C_{TOL}\right)\varphi$

Catalyst deactivation function $\boldsymbol{\varphi}$ for the reactant conversion model **φ = exp(-λx)**

Reaction rate constant $\boldsymbol{k_i}$ is a function of temperature (Arrhenius equation): $k_i = k_{i0}\exp\left[-\frac{E_i}{R}\left(\frac{1}{T}-\frac{1}{T_0}\right)\right]$

where
C_i = molar concentration of i, V = reactor volume, W_c = catalyst mass, and t = reaction time, λ = catalyst deactivation constant, and x = conversion of C9 aromatics, k_{i0} = pre-exponential factor, $\boldsymbol{E_j}$ = activation energy of reaction j.

-22-

The rate equations were reformulated by adding the terms for the fifth reaction.

Comparison of Kinetic Parameters

Reaction	MEB Dealkylation		TMB Disproportionation		Toluene-TMB Transalkylation		TeMB Paring		Conversion of MEB to Xylene		Catalyst Deactivation Function
Kinetic Parameter	k_1 m^3/kgcat.s (x 10^3)	E_1 (kJ/mol)	k_2 m^6/kgcat.s (x 10^3)	E_2 (kJ/mol)	k_3 m^6/kgcat.s (x 10^3)	E_3 (kJ/mol)	k_4 m^3/kgcat.s (x 10^3)	E_4 (kJ/mol)	k_5 m^3/kgcat.s (x 10^3)	E_5 (kJ/mol)	λ
Four (4) Reaction Kinetic Model											
Value	6.01	21.00	0.03	15.05	1.00	10.11	3.45	96.15			0.79
95% CL	0.82	1.31	0.01	2.87	0.15	1.53	0.321	8.36			0.21
Five (5) Reaction Kinetic Model											
Value	1.31	29.25	0.03	16.82	0.06	11.46	2.71	46.33	0.98	19.54	0.41
95% CL	0.06	2.43	0.00	1.56	0.01	1.94	0.15	3.83	0.04	2.41	0.15

-23-

Here are the estimated kinetic parameters for the five reactions and the catalyst deactivation function.

Activation energies for the corresponding reactions are comparable.

Lowest activation energy is for Toluene-TMB transalkylation which is the main xylene producing reaction.

Highest activation energy is for TeMB paring reaction which involve the bulky C10 molecule.

Model Evaluation: Parity Plots and Correlation Coefficients

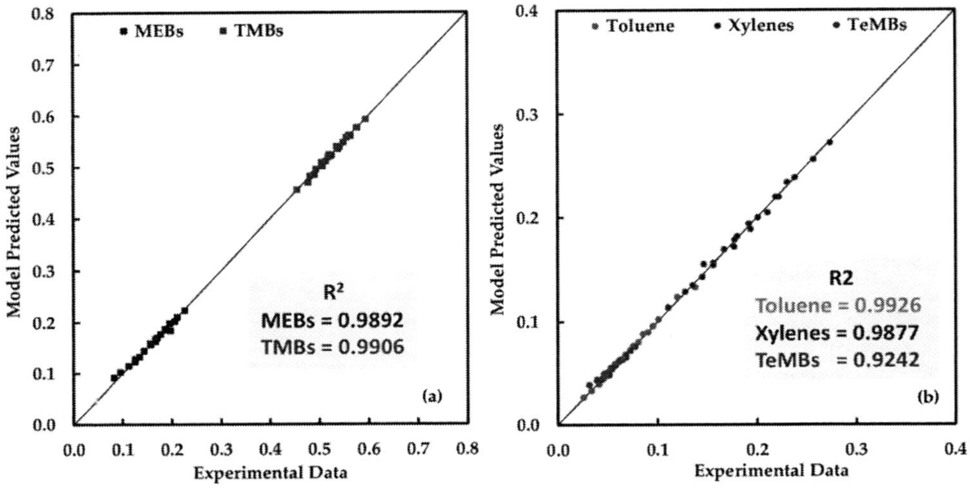

Comparison of R^2 Values		
Species	Four Reaction Network	Five Reaction Network
MEBs	0.9910	0.9892
TMBs	0.9936	0.9906
Toluene	0.9699	0.9926
Xylenes	0.9950	0.9877
TeMBs	0.6446	0.9242

Improvement in predicted product composition by 5-reaction model

-24-

Plot (a) shows the parity plot for the unconverted feed components which are MEBs and TMBs. The model seems to predict their composition quite well. Both R2 values are quite acceptable.

Plot(b) shows the parity plot for the products which are toluene, xylenes, and TeMBs. There is clear improvement in the predicted composition of the three product.

All correlation coefficients are in the acceptable range.

Comparison of Activation Energies with Reported Values

Source of Data	Catalyst	MEB Dealkylation	TMB Disproportionation	Toluene-TMB Transalkylation	TeMB Paring	Conversion of MEB to Xylene
4-Rx Model	Beta-MCM41 Composite	21.0	15.1	10.1	96.2	-
5-Rx Model		29.3	16.8	11.5	46.3	19.5
Ref-1	Mordenite	-	12.2	7.2	-	-
Ref-2	Mordenite (180) + ZSM5 (27)	34.9	34.2	17.8	83.4	-
	Mordenite (18) + ZSM5 (27)	42.1	42.4	27.6	71.7	-
Ref-3	Mordenite (18)	25.5	57.2	6.8	-	-
	Beta (25)	14.3	14.3	11.2	-	-
	Beta (25)	-	-	-	-	12.6
Ref-4	Plant Data	39.5	50.0	31.2	60.6	-

Ref-1: Waziri et al., Ind. Eng. Chem. Res., 49 (2010) 6376
Ref-2: Al-Mubaiyedh et al., Chem. Eng. Res. Des., 90 (2012) 1943
Ref-3: Ali et al., Chem. Eng. Res. Des., 91 (2013) 2601
Ref-4: Hamedi et al., J. Taiwan Inst. Chem. Eng., 48 (2015) 56

-25-

The activation energies obtained by this study are compared well with the reported values.

In all the studies, toluene-TMB transalkylation reaction showed lowest activation energy while TeMB paring reaction showed highest activation energy.

The type of zeolite as well as the silica-alumina ratio (thus the acidity) plays an important in the kinetics of different reactions.

Concluding Remarks

- The metal-loaded bifunctional composite catalyst with higher mesoporous volume than the parent zeolite facilitated better access to large reactant molecules.
- For the zeolite-beta based composite catalyst, five-reaction network is more representative for the overall conversion of heavy reformate as shown by kinetic modeling results.
- Kinetic modeling results also show that the transalkylation reaction was significantly faster than the disproportionation reaction indicating that the transfer of methyl group from TMB to toluene is a preferred route under the reaction conditions studied.
- The high rate of formation of xylenes directly from MEBs (Reaction 5) compared to the indirect route (thru Reactions 1 and 3) corroborate the earlier finding that the topology of catalyst play a major role in the conversion of heavy reformate.

-26-

- The metal-loaded bifunctional composite catalyst with higher mesoporous volume than the parent zeolite facilitated better access to large reactant molecules.

- For the zeolite-beta based composite catalyst, five-reaction network is more representative for the overall conversion of heavy reformate as shown by kinetic modeling results.

- Kinetic modeling results also show that the transalkylation reaction was significantly faster than the disproportionation reaction indicating that the transfer of methyl group from TMB to toluene is a preferred route under the reaction conditions studied.

- The high rate of formation of xylenes directly from MEBs (Reaction 5) compared to the indirect route (thru Reactions 1 and 3) corroborate the earlier finding that the topology of catalyst play a major role in the conversion of heavy reformate.

With this I conclude my presentation and I will be glad to respond to our questions or comments. Thank you all for your patience.

thank you
danke
謝謝
ngiyabonga
شكراً جزيلاً
teşekkür ederim
спасибо
gracias
dank je
mochchakkeram
dziękuję
go raibh maith agat
obrigado
sukriya
kop khun krap
grazie
arigatō
dakujem
мерси
merci
감사합니다
ευχαριστώ
ありがとう
terima kasih
takk
Баярлалаа
рахмат
хвала

CO_2 Assisted Oxidative Dehydrogenation of Propane to Propylene over Fluidizable MoO_3/La_2O_3-γAl_2O_3 Catalysts

Mohammad M. Hossain [a,b*]

[a] Department of Chemical Engineering, [b]Center for Refining and Advanced Chemicals,

King Fahd University of Petroleum and Minerals, Dhahran, Saudi Arabia.

[b] Center of Research Excellence in Nanotechnology, King Fahd University of Petroleum and Minerals, Dhahran, Saudi Arabia.

Abstract

This study investigated the catalytic performance of a novel fluidizable MoO_3 supported on La_2O_3-γAl_2O_3 composite for oxidative dehydrogenation (ODH) of propane to propylene with and without CO_2. The La_2O_3-γAl_2O_3 composites were synthesized by a coprecipitation method, while the active MoO_3 was deposited by an incipient wetness impregnation approach. The ODH of propane experiments were developed in a fluidized CREC Riser Simulator by feeding propane with or without CO_2. The performance of the synthesized catalysts was found to be a function of the La_2O_3 content of the support. In general, the propane conversion declined slightly, while propylene selectivity increased with the increasing La_2O_3 content. It was hypothesized that the presence of basic La_2O_3 reduced the acidity and the non-selective sites of γ-Al_2O_3, which were responsible for deep oxidation of propylene and propane. Among the studied catalysts, MoO_3/La_2O_3-$\gamma Al_2O_3$1:2 displayed the highest propylene yield of 28.2 % at 600 °C in the absence of CO_2. However, a superior yield of 35.2% was achieved over the same catalyst by the introduction of CO_2-admixture in the feed at 550 °C.

Key words: Oxidative dehydrogenation; propane; propylene; mild oxidant; CO_2

1. Introduction

In recent years, the global demand for light olefins (i.e. ethylene, propylene and butenes) has been increasing significantly. Among the short chain olefins, propylene is the most important feedstock in the petrochemical and chemical industries for the production of polypropylene, acrylonitrile, cumen, phonol, iso-propylic alcohol, propylene oxide and several other variety of propylene derivatives [1][2][3]. The projected annual global propylene demand by the year of 2030 is approximately 160 million metric tons [4][5]. The estimation also shows that there is a gap between propylene demand and supply, which is likely to grow further in the near future [6][7]. Over 80 % of the current world propylene supply is obtained as secondary product of the fluid catalytic cracking (FCC) of vacuum gas oil and steam cracking of naphtha and other hydrocarbons.

The remaining 20% of the propylene supply is mainly derived from the on-purpose processes such as dehydrogenation of propane, oxidative dehydrogenation of propane, olefins metathesis and methanol-to-propylene [8][9]. The dynamics of propylene supply from the FCC processes is related to the price of gasoline. As the demand for gasoline decreases, the refineries are forced to cut down production, which affects the propylene supply [10]. Severe coke formation is another issue related to FCC based propylene production. On the other hand, steam cracking is an energy intensive process, which accounts for almost 40% of the total annual energy consumption in the petrochemical industries. The steam cracking process also suffers from coke formation and it significantly contributes to the emission of CO_2 into the atmosphere [11][12][13]. In order to address the after mentioned issues, several innovative technologies have been explored to produce propylene from alternative sources [14].

The present study investigated the MoO_3/La_2O_3-γAl_2O_3 catalysts in oxidative dehydrogenation of propane using CO_2 as a mild oxidant. The catalyst support system with varying ratio of $La_2O_3/\gamma Al_2O_3$ was synthesized by the coprecipitation technique. The active MoO_3 phase was impregnated on the mixed $La_2O_3/\gamma Al_2O_3$ support by an incipient wetness approach. The CO_2 assisted ODH of propane was carried out in a fluidized CREC Riser Simulator at atmospheric pressure and at varying temperature (525-600°C). The stability of the catalyst, effect temperature, effect of CO_2, effect of $La_2O_3/\gamma Al_2O_3$ ratios have been investigated.

2. Experimental

2.1. Catalyst preparation

The La_2O_3-γAl_2O_3 composite supports were prepared by a coprecipitation technique, with $La_2O_3/\gamma Al_2O_3$ ratio of 1:3, 1:2 and 1:1. The MoO_3 were deposited on the La_2O_3-γAl_2O_3 support by an incipient wetness impregnation technique. In all catalysts, the Mo loading was maintained at 10 wt % and the synthesized catalysts are identified as (i) $MoO_3/\gamma Al_2O_3$, (ii) MoO_3/La_2O_3-$\gamma Al_2O_3$1:3, (iii) MoO_3/La_2O_3-$\gamma Al_2O_3$1:2 and (iv) MoO_3/La_2O_3-γAl_2O_3 1:1.

2.2. Catalyst Evaluation

The CO_2 assisted ODH of propane experiments were carried out in a batch scale fluidized CREC Riser Simulator (CREC- Chemical Reaction Engineering Center). All the experiments conducted using same 0.2 g of catalyst. The products weres analyzed by an online GC (Agilent 7890A) equipped with both TCD and FID analyzer. The product analysis data was further processed to calculate propane conversion ($X_{C_3H_8}$), products selectivity (S_i) and yields(Y_i), according to the following equations:

$$\text{Propane conversion, } X_{C_3H_8}(\%) = \frac{\sum_j z_j n_j}{3n_{propane} + \sum_j z_j n_j} \times 100 \quad (1)$$

$$\text{Selectivity to a product, } S_j(\%) = \frac{z_j n_j}{\sum_j z_j n_j} \times 100 \qquad (2)$$

where, z_j and n_j are the number of atoms of carbon and moles of gaseous carbon containing product j, respectively, $n_{propane}$ is the mole of unconverted propane in the product stream.

3. Results and Discussion

Figure 1 compares the propane conversion at various reaction temperature with and without the addition of CO_2 in the feed. Interestingly, the propane conversions significantly increased when CO_2 was co-fed along with propane. The superior propane conversion in the presence of CO_2 was attributed to the direct interaction between propane and CO_2 to form propylene and CO and water as by-products. The reverse water-gas shift reaction may also be a contributing factor to the high propane conversion as well as olefin selectivity as seen in Figure 1, 2 and 3. In addition, the injected CO_2 could also have facilitated the insitu catalyst re-oxidation, which eventually contributed to the oxidative dehydrogenation of propane (further discussion to follow).

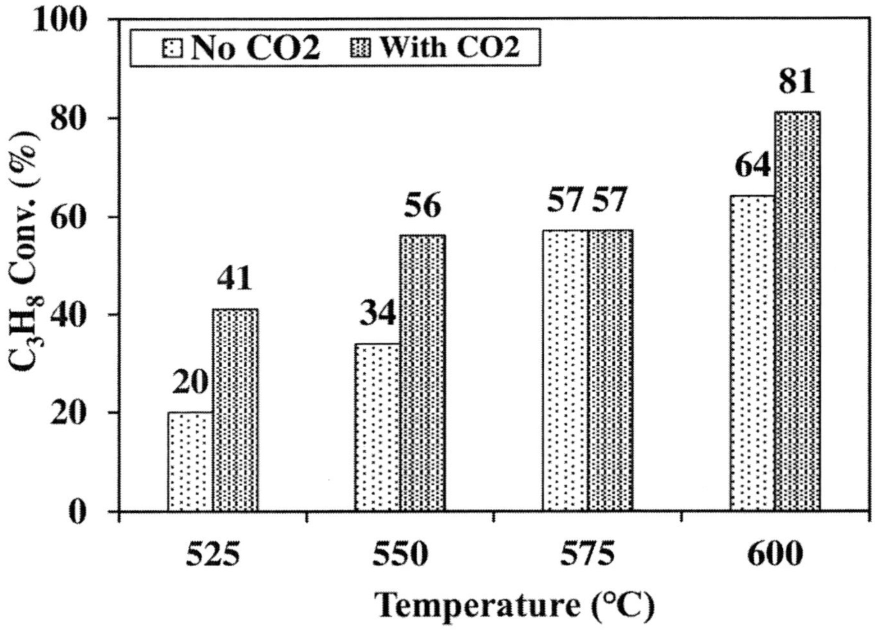

Figure 1: Effect of CO_2 on propane conversion with MoO_3/La_2O_3-γAl_2O_3 1:2 ($W_{cat.}$: 0.2 g, Feed: 2.0 ml C_3H_8, 4.0 ml CO_2, time:30 s).

Concerning the desired product selectivity, one can see that both propylene and ethylene selectivity were significantly higher with CO_2 co-fed experiments than the experiments without CO_2, except at 600 °C (Figure 2 and Figure 3). The addition of CO_2 in feed could have blocked the non-selective electrophilic sites, which contributed to the formation of undesired products. The excess CO_2 in the reaction mixture also influenced the thermodynamic equilibrium of the complete oxidation reactions to minimize further oxidation of propylene to CO_2. On the contrary, at 600 °C deep oxidation dominated the oxidative dehydrogenation, consequently the selectivity of propylene and ethylene dropped.

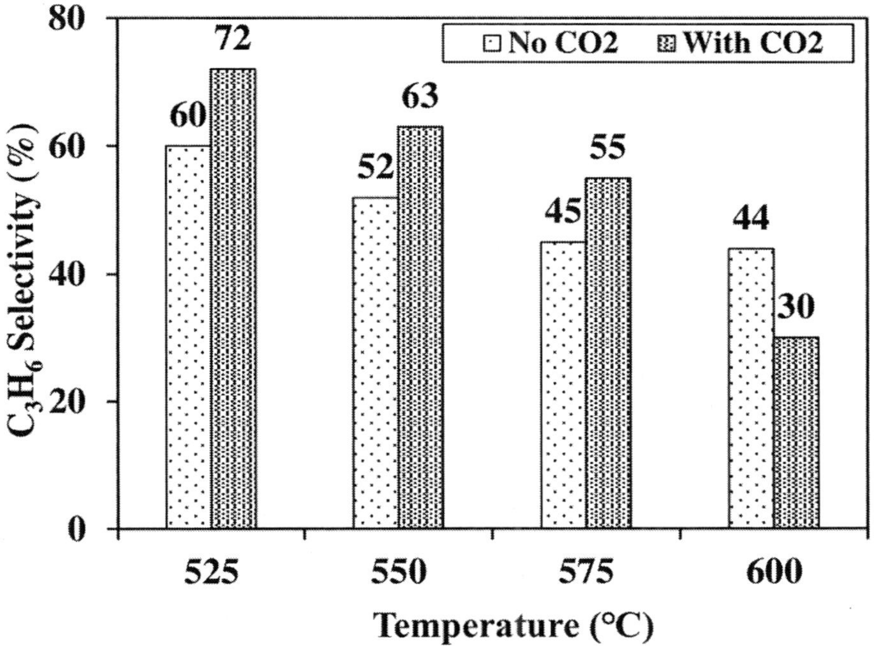

Figure 2: Effect of CO_2 on propylene selectivity with MoO_3/La_2O_3-γAl_2O_3 1:2 (W_{cat}.: 0.2 g, Feed: 2.0 ml C_3H_8, 4.0 ml CO_2, time: 30 s).

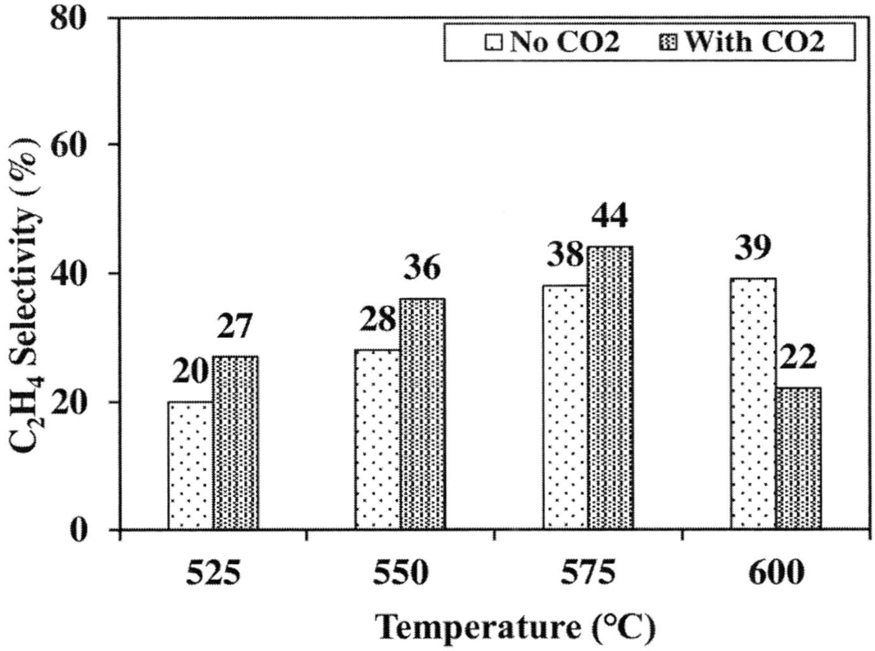

Figure 3: Ethylene selectivity with CO_2 (W_{cat}.: 0.2 g, Feed: 2.0 ml C_3H_8, 4.0 ml CO_2, , time: 30s).

3.1. Successive Propane Injections – Without and With CO_2

Figure 4 presents the propane conversion and propylene selectivity for seven successive propane injections without catalyst regeneration by air circulation (as done in the previous experiments). The first five consecutive runs were conducted by injecting propane only, without catalyst regeneration, while in the last two experiments, propane was co-fed with CO_2 at propane/CO_2 ratios of 1:2 and 1:3, respectively. The highest propane conversion of 42% was obtained in the first injection, as the catalyst was at fully oxidized state. Following the first injection, the propane conversion gradually decreased to 33% within five injections. The initial high propane conversion was ascribed to the availability of weakly bonded surface lattice oxygen of the fully oxidized catalyst, which contributed to the activation of C-H bond of propane to give propylene and carbon oxides. The decrease in propane conversion during the following injection was due to the depletion of the surface and lattice oxygen of the catalyst. On the other hand, the propylene selectivity increased from an initial 33% to 42% during the successive propane injections. The increase in propane selectivity in the consecutive propane injection was attributed to major contribution of the lattice oxygen, given the surface oxygen gives complete oxidation products. In the first injection, the catalyst was fully oxidized and it contained weakly bonded surface oxygen, which was more selective to deep oxidation of propane and propylene. In the following injections, the depletion of the non-selective lattice oxygen helped to increase the propylene selectivity. On the other hand, the nucleophilic nature of the strongly bonded lattice oxygen contributed to selective oxidation of hydrocarbon to desired olefin products.

During the last two injections (i.e. runs 6 and 7), propane was co-fed with CO_2 to demonstrate the effects of CO_2. One can clearly see that the introduction of CO_2 increased the propane conversion from 33 % to 35 % and propylene selectivity from 42 % to 45 %. After the seventh injection, the propane conversion and propylene selectivity further increased to 36% and 48%, respectively because of the increased CO_2 concentration in the reaction mixture. This is an indication that CO_2 has promotional effects on propane ODH as it influences the thermodynamic equilibrium of the complete oxidation reactions to minimize CO_2 formation, CO_2 reacts directly with propane over the catalyst surface to produce propylene, and re-oxidizes the oxygen depleted catalysts [104][35][105]. Thus, CO_2 may be acting as an adsorbate, which covers the non-olefin selective sites of the catalyst to minimize oxidation of propylene. Again, the increase in propane conversion in the presence of CO_2 may also be partly attributed to the reverse water-gas shift reaction (Eq. 13), which consumes H_2 generated from direct propane dehydrogenation (Eq. 12) to enhance the propylene selectivity as the forward reaction is favored.

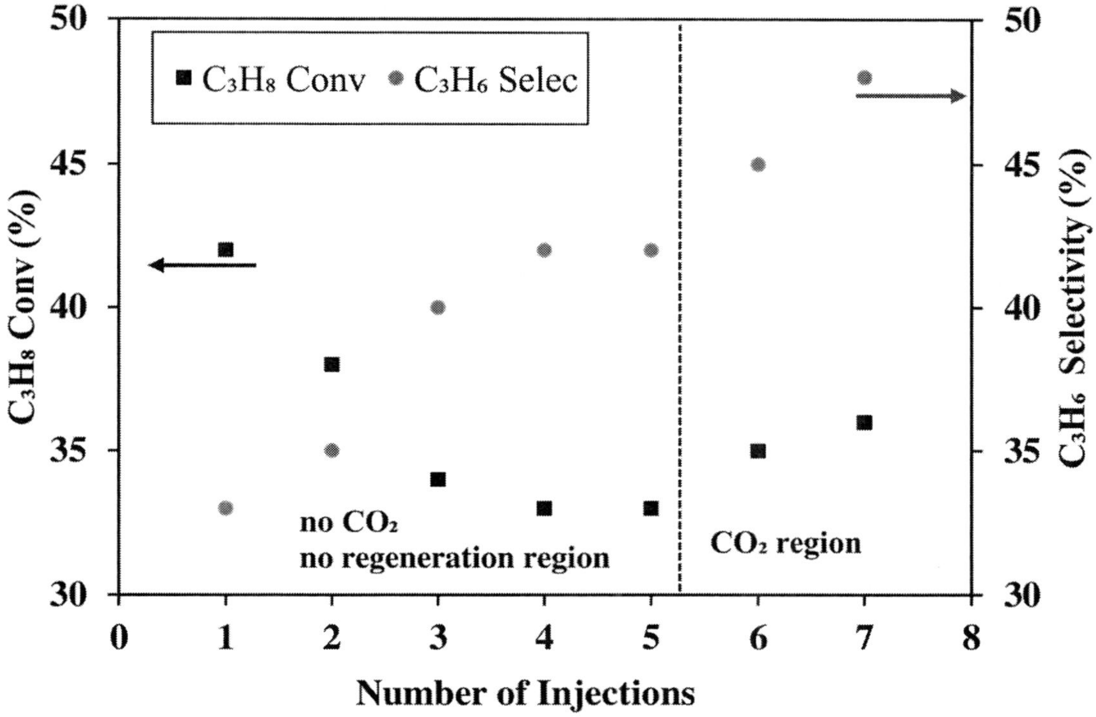

Figure 4: Propane conversion and selectivity of propylene over $MoO_3/\gamma\text{-}Al_2O_3$ in consecutive propane injection without catalyst regeneration. The first 5 consecutive injections are propane only, the 6th and 7th are C_3H_8 with CO_2 in ratio 1:2 and 1:3 respectively.

4. Conclusion

The presence of CO_2 in the feed showed considerable influence on the catalytic activity as well as the propylene selectivity which may be due to its poisoning effect by blocking the non-olefin selective sites available on the catalyst. This study also demonstrates that CO_2 is capable of partially re-oxidizing reduced catalyst by replacing the selective lattice oxygen which improves olefin selectivity. Finally, the synthesized catalysts displayed high olefins (ethylene and propylene) selectivity and significant stability over a number of consecutive reactions cycles and could be used in an industrial fluidized bed reactor.

Acknowledgements

The author(s) would like to acknowledge the support provided by the Deanship of Scientific Research (DSR) at King Fahd University of Petroleum & Minerals (KFUPM) for funding this work through project No. DF191028.

References

[1] J. Wang, F. Zhang, W. Hua, Y. Yue, Z. Gao, Dehydrogenation of propane over MWW-type zeolites supported gallium oxide, Catal. Commun. 18 (2012) 63–67. https://doi.org/10.1016/J.CATCOM.2011.11.023.

[2] B. Li, Z. Xu, F. Jing, S. Luo, W. Chu, Facile one-pot synthesized ordered mesoporous Mg-SBA-15 supported PtSn catalysts for propane dehydrogenation, Appl. Catal. A Gen. 533 (2017) 17–27. https://doi.org/10.1016/j.apcata.2016.12.026.

[3] M. Fujiwara, N. Mimura, O. Sato, A. Yamaguchi, Surface modification of H-ZSM-5 with organo-disilane compound for propylene production from dimethyl ether, Microporous Mesoporous Mater. 280 (2019) 219–226. https://doi.org/10.1016/j.micromeso.2019.02.005.

[4] E. Nowicka, C. Reece, S.M. Althahban, K.M.H. Mohammed, S.A. Kondrat, D.J. Morgan, Q. He, D.J. Willock, S. Golunski, C.J. Kiely, G.J. Hutchings, Elucidating the Role of CO $_2$ in the Soft Oxidative Dehydrogenation of Propane over Ceria-Based Catalysts, ACS Catal. 8 (2018) 3454–3468. https://doi.org/10.1021/acscatal.7b03805.

[5] Y.-K. Park, C.W. Lee, N.Y. Kang, W.C. Choi, S. Choi, S.H. Oh, D.S. Park, Catalytic Cracking of Lower-Valued Hydrocarbons for Producing Light Olefins, Catal. Surv. from Asia. 14 (2010) 75–84. https://doi.org/10.1007/s10563-010-9089-1.

[6] M. Sarker, D.K. Yoo, S. Lee, T.-W. Kim, C.-U. Kim, S.H. Jhung, Conversion of Y into SSZ-13 zeolite, in the absence of extra silica, alumina and seed crystals, with N,N,N-dimethylethylcyclohexylammonium bromide, and application of the SSZ-13 zeolite in the propylene production from ethylene, Catal. Today. (2019). https://doi.org/10.1016/j.cattod.2019.12.010.

[7] A.I. Hussain, A.M. Aitani, M. Kubů, Jıє. Čejka, S. Al-Khattaf, Catalytic cracking of Arabian Light VGO over novel zeolites as FCC catalyst additives for maximizing propylene yield, Fuel. 167 (2016) 226–239. https://doi.org/10.1016/j.fuel.2015.11.065.

[8] M.F. Alotibi, B.A. Alshammari, M.H. Alotaibi, F.M. Alotaibi, S. Alshihri, R.M. Navarro, J.L.G. Fierro, ZSM-5 Zeolite Based Additive in FCC Process: A Review on Modifications for Improving Propylene Production, Catal. Surv. from Asia. 24 (2020) 1–10. https://doi.org/10.1007/s10563-019-09285-1.

[9] A. Akah, J. Williams, M. Ghrami, An Overview of Light Olefins Production via Steam Enhanced Catalytic Cracking, Catal. Surv. from Asia. (2019). https://doi.org/10.1007/s10563-019-09280-6.

[10] A. Agarwal, D. Sengupta, M. El-Halwagi, Sustainable Process Design Approach for On-Purpose Propylene Production and Intensification, (2017).

https://doi.org/10.1021/acssuschemeng.7b03854.

[11] M. Numan, T. Kim, C. Jo, S.-E. Park, Ethane Dehydrogenation with CO2 as a soft oxidant over a Cr-TUD-1 catalyst, J. CO2 Util. 39 (2020) 101184. https://doi.org/10.1016/j.jcou.2020.101184.

[12] Z.-F. Han, X.-L. Xue, J.-M. Wu, W.-Z. Lang, Y.-J. Guo, Preparation and catalytic properties of mesoporous nV-MCM-41 for propane oxidative dehydrogenation in the presence of CO2, Chinese J. Catal. 39 (2018) 1099–1109. https://doi.org/10.1016/S1872-2067(18)63048-7.

[13] P.T. Bai, V. Manokaran, P.S. Saiprasad, S. Srinath, Studies on Heat and Mass Transfer Limitations in Oxidative Dehydrogenation of Ethane over Cr2O3 /Al2O3 Catalyst, in: Procedia Eng., Elsevier Ltd, 2015: pp. 1338–1345. https://doi.org/10.1016/j.proeng.2015.11.492.

[14] F. Zhang, R. Wu, Y. Yue, W. Yang, S. Gu, C. Miao, W. Hua, Z. Gao, Chromium oxide supported on ZSM-5 as a novel efficient catalyst for dehydrogenation of propane with CO2, Microporous Mesoporous Mater. 145 (2011) 194–199. https://doi.org/10.1016/J.MICROMESO.2011.05.021.

136a Natural Gas Utilization and Resilience Study for Simultaneous Production of Electricity, Nitrogen Fertilizer and Salicylic Acid with Inherent CO_2 Sequestration

Ying Liu, Viral Modi, Sujing Wang, Qiang Xu

The increasing global human population, rapid growth of energy and food demand, and the escalating concern of global CO_2 emissions will continuously bring great challenges to the world. As one of the most abundant and available alternative energy/material resources, natural gas should be comprehensively utilized to help mitigate these concerns. In this paper, a material-heat-power integrated conceptual design of an industrial complex for comprehensive natural gas utilization has been proposed, modeled, and examined. It includes five subsystems (termed A^3US): (i) an air separation unit (ASU); (ii) an Allam-cycle power plant (ACPP); (iii) an ammonia plant (APP); (iv) a urea production plant (UPP), and (v) a salicylic acid plant (SAP). The complex employs natural gas and air as the main feedstock. Through optimal integraton of their energy and material streams among five subsystems, the A^3US complex will simultaneously produce multiple products including net power, urea, salicylic acid, pipeline-ready CO_2, pure N_2, and steam effectively and efficiently. Meanwhile, the A^3US complex completely has no NOx emissions and barely has CO_2 emissions. The developed A^3US complex has been virtually demonstrated by large-scale rigorous modeling and simulations. Based on the simulation results, two resilience indexes have also been developed to quantify the system resilience of the developed A^3US complex: (i) the emission resilience index gauging the ratio of CO_2 utilization and sequestration rates to the total CO_2 generation rate and (ii) the economic resilience index measuring the ratio of compound product profit to that from average product markets. These two resilience indexes help highlight the sailent environmental and economic benefits of the A^3US complex.

136b Linde's EDHOX™ Technology: Commercial Ethylene and Acetic Acid Production without Direct CO_2 Emissions

Gunther Kracker, Ricardo Bermejo

Linde's EDHOX™ technology is the first commercial process to produce ethylene (polymer grade) by the oxidative dehydrogenation of ethane, offering significantly lower specific energy consumption and lower CO_2 emissions than conventional ethane-steam cracking (E-SC). In addition, the EDHOX™ technology co-produces acetic acid (glacial purity) and is the ideal choice for downstream processes requiring both ethylene and acetic acid, such as VAM (Vinyl Acetate Monomer), EVA (Ethylene Vinyl Acetate) copolymer, PVOH (Polyvinyl Alcohol products), PTA (Terephthalic Acid), ethyl acetate and similar derivatives.

The Linde EDHOX™ technology provides a wide flexibility range of ethylene to acetic acid production (2.5-4.5 t/t) and an overall selectivity towards these products above 93% under safe and reliable oxidative conditions. This is achieved with a multitubular reactor at low pressure and moderate operating temperature (300-350°C), which can be materialized in a wide range of production capacities up to world scale ethylene plant capacities. Furthermore, its oxidative route is exothermic and consequently does not require additional heat input by firing or electricity. The high selectivity to ethylene and acetic acid is achieved with a solid catalyst, producing, in turn, less by-products than conventional E-SC and simplifying product separation and purification. CO_2 is only formed as a reaction by-product and captured as part of the EDHOX™ process prior to the ethylene purification. These unique features of the EDHOX™ technology result in more than three times lower specific energy consumption than conventional E-SC. With utilization or storage of the captured CO_2 and the usage of renewable power the Linde EDHOX™ technology can effortlessly attain "Zero Scope 1 & 2 Emissions".

The EDHOX™ technology performance has been verified in a demonstration plant under commercial conditions at Linde's Pullach (Munich) Engineering center in Germany, with a commercial size reactor tube, separation-purification and closed-loop operation to recycle non-converted ethane. Linde is currently working on several early phase projects for the first commercial scale application of the EDHOX™ technology.

The paper describes in detail the Linde EDHOX™ technology and highlights its "Zero Scope 1 & 2 Emissions" features. It consequently focuses on its significant economic advantages in comparison to conventional E-SC. This includes up to 20% lower CAPEX, 40% lower ethylene production costs per ton of ethylene and significantly higher net present value (NPV) within historic feed and product prices, regardless of CO_2 credits.

73b Three Disruptive Technologies Coming Soon to Manufacturing and Process Control

Tom Badgwell, R Donald Bartusiak

We are currently in the midst of a fourth industrial revolution (Industry 4.0 [1]), involving the large-scale automation of traditional manufacturing and industrial practices, made possible by recent developments in mathematical algorithms, computer hardware, and internet connectivity (Industrial Internet of Things (IIoT) [2]). While much of this work can be considered evolutionary in nature, in this presentation we highlight three emerging technologies that appear to be truly disruptive; that is, they are likely to have such a large impact that they will change the way theoreticians and practitioners think about and accomplish manufacturing and process control. These technologies are Economic Model Predictive Control (EMPC), Deep Reinforcement Learning (DRL), and Open Process Automation (OPA).

Economic MPC (EMPC) is a relatively new technology that combines economic optimization with Model Predictive Control [3], two functions that are traditionally implemented separately. While the theory was worked out a few years ago [4], EMPC applications have only begun to appear recently. Professor Jim Rawlings and co-workers, for example, presented a successful EMPC implementation for the Stanford University campus heating and cooling system [5]. Recent theoretical work has shown that scheduling problems, which are usually approached from the point of view of static optimization, can also be considered as a special case of closed-loop EMPC [6]. This unification of closed-loop scheduling, economic optimization, and dynamic control has shed new light on such problems as rescheduling in the face of disturbances, and provides academics with a completely new framework for viewing and analyzing scheduling problems. Practitioners now have, for the first time, the hope of combining three disparate levels in the traditional control hierarchy into a single, harmonious layer.

Reinforcement Learning (RL) is a Machine Learning (ML) technology in which a computer agent learns, through trial and error, the best way to accomplish a particular task [7]. Deep Learning (DL) is a technology in which neural networks with a large number of intermediate layers are used to model relationships [8]. Using DL to parametrize the policy and value function of an RL agent leads to Deep Reinforcement Learning (DRL) technology, which allows an agent to achieve superhuman performance for some tasks. In 2017, for example, a DRL agent named AlphaGo soundly defeated the reigning world champion Go player [9]. Applications of this technology to manufacturing and process control systems are currently under study [10]. It is likely that DRL will not replace currently successful control algorithms such as PID and MPC, but will rather takeover some of the mundane tasks that humans perform to manage automation and control systems. For example, it appears that a DRL agent can learn how to tune PID loops effectively [11]. Other possibilities include advising operators during transient and upset conditions, mitigating disturbances such as weather events, and detecting and mitigating unsafe operations [10].

The industrial automation marketplace, comprised of Distributed Control System (DCS), Programmable Logic Controller (PLC), and Supervisory Control and Data Acquisition (SCADA) technology offerings, will soon experience a historic, game-changing disruption with the emergence of Open Process Automation (OPA) technology. Manufacturers, whose innovations have been constrained for decades by the limitations of closed, proprietary systems, will soon experience the benefits of open, interoperable, resilient, secure-by-design automation systems, made possible by the development of the consensus-based Open Process Automation Standard (O-PAS) by the Open

Process Automation Forum (OPAF) [12]. Once O-PAS certified automation systems become widespread, vendors will see the market for their products and services expand significantly as the visions of I4.0 and the IIoT are realized. Academics and technology developers will see more opportunities to test their solutions as it becomes easier to deploy them. Dr. Don Bartusiak, co-director of OPAF, summarizes their progress to date in a paper published recently in Control Engineering Practice [12].

References

[1] Y Liao, F Deschamps, EdFR Loures, LFP Ramos, "Past, present, and future of Industry 4.0 - a systematic literature review and research agenda proposal", Intl J Production Research, 55 (12), 3609-3629, (2017).

[2] H Boyes, B Hallaq, J Cunningham, T Watson, "The industrial internet of things (IIoT): An analysis framework", Computers in Industry, 101, 1–12, (2018).

[3] SJ Qin, TA Badgwell, "A survey of industrial model predictive control technology", Control Engineering Practice 11 (7), 733-764, (2003).

[4] JB Rawlings, D Angeli, CN Bates, "Fundamentals of economic model predictive control", 51st Conference on Decision and Control, 3851-3861, (2012).

[5] JB Rawlings, NR Patel, MJ Risbeck, CT Maravelias, MJ Wenzel, RD Turney, "Economic MPC and real-time decision making with application to large-scale HVAC energy systems", Computers & Chemical Engineering, 114 (6), 89-98, (2018).

[6] MJ Risbeck, CT Maravelias, JB Rawlings, "Unification of Closed-Loop Scheduling and Control: State-space Formulations, Terminal Constraints, and Nominal Theoretical Properties", Computers & Chemical Engineering, 129 (10), (2019).

[7] RS Sutton, AG Barto, "Reinforcement Learning – An Introduction", The MIT Press, (2018).

[8] I Goodfellow, Y Bengio, A Courville, "Deep Learning", The MIT Press, (2016).

[9] D Silver, A Huang, CJ Maddison, A Guez, L Sifre, G Van Den Driessche, J Schrittwieser, I. Antonoglou, V Panneershelvam, M Lanctot, "Mastering the game of Go with deep neural networks and tree search", Nature 529, 484–489 (2017).

[10] J Shin, TA Badgwell, KH Liu, JH Lee, "Reinforcement Learning – Overview of recent progress and implications for process control", Computers & Chemical Engineering, 127, 282-294 (2019).

[11] TA Badgwell, KH Liu, NA Subrahmanya, WD Liu, MH Kovalski, "Adaptive PID Controller Tuning via Deep Reinforcement Learning", U.S. patent 1095073, granted February 9, 2021.

[12] RD Bartusiak, S Bitar, DL DeBari, BG Houk, D Stevens, B Fitzpatrick, P Sloan, "Open Process Automation: A Standards-Based, Open, Secure, Interoperable Process Control Architecture", Control Engineering Practice 121, 105034, (2022).

73c Perform Better Planning and Scheduling of VCM Furnace Decoke-Shutdowns with Enhanced Capability of Load Balancing, through Predictive Modeling

Himanshu Malaiya

Cracking of EDC in furnaces produces coke as a byproduct, this coke layers retards heat transfer and increases tube metal temperature and pressure drop across the coil. The accumulation of deposits on the surface of the coil necessitates periodic removal of this coke by steam-air decoking. Thus after a few months of operation, furnace shut down is taken for decoking of the furnace. In cracking furnaces, it is difficult to estimate the remaining run-length of the furnaces and is generally determined intuitively with the help of process parameters trends, furnace coil pressure drop and TMT. This makes it difficult to plan and schedule decoke shutdowns for these furnaces. Any unplanned decoke-shutdown, will ultimately lead to productivity losses. In addition, absence of coke monitoring, load balancing is performed intuitively compounding the issue of unplanned decokes.

For a VCM unit in USA, a forecasting model was developed to predict the furnace run-length based on critical operating parameters. This model is trained on historical data and is continuously fed with live data from the plant. The model is a dynamically auto-tuned model making it receptive to changes in plant operating conditions. Therefore, in case of any sudden increase in slope of pressure drop, it will immediately alert the operator to take actions to control the coking rate.

This model is supplemented with a deviation tracker which will also help the operator identify the coil and tubes where coking rate is higher, so that appropriate actions can be taken by adjusting the burners near that tube.

To make the overall application more utilizable in decision making, a "What-if" feature was built to help predict the furnace run-length at different feed rates. For example, it can indicate that the furnace run-length will increase/decrease by 10 days if the feed is decreased/increased by 5% today. This will help in better production planning of the furnaces and will help in aligning the furnace end-of run dates with the planned decoking dates.

Application Utilization:

1. **Monitor** furnace coking rates and implement operational changes to control the coking rates and achieve desired run length.
2. **Perform** "What-if" provision to predict future furnace performance at different feed rates, for better planning and scheduling of furnace operation and decoke cycles.
3. **Perform** informed load balancing of the furnaces, with combined insights from above evaluations.

Application Benefits:

- The model has been made online in the discussed unit few months ago, and is performing as expected.
- Quantitively, potential benefits of implementing this model in the unit are estimated to be ~ $0.5 million/annum.
- Qualitatively, augmenting the intelligence of field/panel engineers with such a tool will help in pro-active and continuous improvement of furnace operation leading to better asset utilization without requiring any major CAPEX with enhanced reliability and financial benefits.

91a Engineered DMC Models with Incremental Mass-Balance Constraints

Vivek Dabholkar

Unlike Equation Based Optimizer, DMC Controller Models are inherently Empirical based models. Therefore, there are no certainties that incremental mass balance(s) is/are honored. Therefore, models must be engineered to enforce Incremental Mass Balance Conservation. In this presentation, we will present two examples with increasing complexity to illustrate the point. Frequently encountered first case where the Distillation Tower Bottom Section Tray Temperature Setpoint is a Manipulated Variable and then perhaps less frequently encountered second case where the Reboiler Flow Setpoint is a Manipulated Variable.

91b Heater Pass Balancing using MPC on a Redundant DCS node

Lanny Gibson, Doug Jackson, Logan Moss, Allan Kern, Tyler Smith

A multivariable controller to balance heat distribution by adjusting individual heater pass flows, while maintaining upstream level control and managing other process constraints was implemented in a redundant DCS node. The controller overcame various instrumentation issues including significant valve stiction to reduce heater pass imbalances, skin temperature excursions, and improve upstream level stability across a wide range of process flow rates. The controller was designed and implemented using Lin and Associates' Model-less Multivariable Controller, Deltum-XMC, installed in the Honeywell Experion C300 control node, and commissioned in less than a calendar-week without step-tests or process modeling. The controller is running at a 2-second in the C300 which is a redundant control node.

91c Adaptive Process Control Using Chemical Engineering Fundamentals

Keith Lehman

One weakness of traditional regulatory and advanced process control (APC) applications is their assumption that the underlying process is linear, i.e. it always behaves in exactly the same way. Non-linear control (NLC) technology can be applied in extreme cases, but usually with significantly higher cost and difficulty level. In many cases, traditional linear control applications can be configured to automatically adjust, or adapt, to changing conditions using chemical engineering fundamentals such as hydraulics, material balances and thermodynamics. This presentation will discuss the need for adaptive process controls, including several practical examples that use basic chemical engineering principles to improve control performance over a range of operating conditions, and some important requirements and limitations for implementing adaptive process control applications.

152a A Data-Driven Framework for Monitoring the Operation of Gas-to-Liquid (GTL) Plants

Mohamed Nounou, Hazem Nounou, Nimir Elbashir, Nour Basha, Radhia Fezai, Byanne Malluhi, Gasim Ibrahim

Gas-to-Liquid (GTL) processes widely are used to produce chemicals and fuels from natural gas [1]. Proper operation of GTL plants requires tight monitoring of their operating conditions. Large amounts of GTL data are continuously collected, which contain important information about the wellness of these plants. In this work, a data-driven framework is developed to monitor the operation of GTL plants. The developed framework consists of three main tasks: modeling, fault detection, and fault identification. Various data-driven techniques are developed to enhance the accuracy of each of these tasks. For example, linear and nonlinear unsupervised modeling techniques that enhance the accuracy of existing methods are developed, which include multiscale Bayesian principal component analysis (MSBPCA) [2], Bayesian Optimized Kernel principal component analysis (BOKPCA) [3, 4], and Bayesian Optimized neural network (BONN) techniques [5]. These techniques provide enhanced prediction accuracy, especially for nonlinear and noisy data. Furthermore, new fault detection techniques are developed, which include PCA-based generalized likelihood ratio (PCA-GLR) test, multiscale PCA-GLR, and interval PCA-GLR [6]. These techniques allow detecting various types of faults, such as shifts in the mean or changes in variance. Finally, extensions of support vector machines (SVM) [7] and random forest (RF) [8] are also explored for enhanced fault identification. The developed modeling, fault detection, and fault identification techniques will be integrated in the proposed framework for monitoring the operation of GTL plants. The effectiveness of such a framework will be validated using simulated examples as well as real data collected from a Fischer-Tropsch setup at the Gas and Fuels Research Center (GFRC) at Texas A&M University at Qatar.

Keywords: process monitoring; data-driven modeling; fault detection; fault identification; gas-to-liquid (GTL) plants

References:

[1] Bao, B., El-Halwagi, M., Elbashir, N., "Simulation, Integration, and Economic Analysis of Gas-to-Liquid Processes", Fuel Processing Technology 91, 703–713, 2010.

[2] Nounou, M., Bakshi, B., Goel, P. and Shen, X., "Bayesian Principal Component Analysis," Journal of Chemometrics, 16, 11, 576–595, 2002.

[3] Kutateladze, V., "The Kernel Trick for Nonlinear Factor Modeling", International Journal of Forecasting, 38, 1, 165–177, 2022.

[4] Victoria, A. and Maragatham, G., "Automatic Tuning of hyperparameters using Bayesian optimization. Evolving Systems, 12(1):217–223, 2021.

[5] K.-L. Du, M. N. S. Swamy, Neural Networks and Statistical Learning, 2nd Ed., Springer-Verlag London Ltd., 2019.

[6] Harrou, F., Nounou, M., Nounou, H., and Madakyaru, M., "Statistical Fault Detection using PCA-based GLR Hypothesis Testing", Journal of Loss Prevention in the Process Industries, 26, 1, 129-139, 2013.

[7] Cortes, C., Vapnik, V., "Support Vector Networks", Machine Learning, 20, 273-297, 1995.

[8] Mansouri, M., Fezai, R., Trabelsi, M., Nounou, H., Nounou, M., and Bouzrara, K., "Reduced Gaussian Process Regression Based Random Forest Approach for Fault Diagnosis of Wind Energy Conversion Systems", IET Renewable Power Generation, 15, 5, 3612-3621, 2021.

152b Effective Use of Hybrid Modeling (ML + Digital-Twin) to Predict and Maintain Uniform Conversion across Furnaces in VCM Plant

Nishant Jaiswal, Shalin Doshi

Hybrid modeling is a combination of fundamental first principle models and AI/ML algorithms to predict important process specific KPI's like Conversion, Selectivity. Generating effective AI/ML models need large amount of "good data", this data can be generated through fundamental models using multi-year historical operating data. AI/ML models can be trained on this data and deployed to predict these KPIs on real-time plant data. These work as a soft sensor providing real-time insight to the predicted parameters/KPIs.

VCM Furnaces are one of the most critical equipment's in VCM manufacturing process. Generally multiple furnaces are operating in parallel and their combined effluent is processed in downstream distillation columns. Based on the total VCM product flow from VCM column the overall plant conversion can be determined (Even this is a challenge for some plants where interconnections with other plants exists). However, it is generally difficult to determine the EDC conversion in individual VCM furnaces. In absence of real-time individual furnace conversion values, one furnace may be operated at higher conversion and thus subjected to higher firing rates while other furnaces may be operating at lower conversion and thus subjected to lower firing rates. Due to this, coking rates and hence run-length will differ across furnaces, leading to earlier decoke-shutdowns and production loss. Therefore, severity balancing in addition to load balancing across furnaces is crucial for VCM operations.

For a VCM unit in USA, a hybrid modelling approach was used for calculating the EDC conversion in VCM furnaces. Using a huge volume of "clean" historical data, a first principle-based model (digital-twin) was developed to estimate the EDC conversion in each furnace, by tuning it to match with furnace downstream operating conditions. This generated data was fed to an ML model to predict conversion on real-time plant data. The concerned unit had two furnaces operating in parallel. The model provides individual and combined conversion of the furnaces as output.

Application Utilization:

1. **Control and achieve** desired EDC conversion in individual furnaces.
2. **Perform** severity tuning across furnaces to ensure uniform conversion.

Application Benefits:

- The model has been made online in the discussed unit few months ago, and is performing as expected.
- Quantitively, potential benefits of implementing this model in the unit are estimated to be ~ $0.8 million/annum.
- Qualitatively, augmenting the intelligence of field/panel engineers with such a tool will help in pro-active and continuous improvement of furnace operation leading to better asset utilization without requiring any major CAPEX.

152c Real-Time Condition Monitoring of Mechanical Seals in Rotating Equipment

Gabriel Young, Adrian Conrad, Katherine Brill-Schuetz

Rotating equipment, such as turbomachinery and pumps, are often the most critical equipment within a production facility. Extended equipment downtime results in production delays, material shortages, and significant financial losses. To mitigate this issue, many production facilities closely monitor such equipment's operating conditions and performance as part of a predictive and preventative maintenance program.

Many options exist for monitoring the health and performance of rotating equipment, but few provide detailed and actionable data specific to the equipment's mechanical seals. Mechanical seals are one of the most sensitive components within a piece of rotating machinery. As such, seal failure and replacement are frequently cited as the cause of unplanned shutdowns of pumps and compressors. Traditional condition monitoring techniques cannot always determine whether a transient event within the process damaged the mechanical seals, or whether the root cause of an alarm is a failed seal. Advances in condition monitoring spurred by Industry 4.0 technologies allow reliability practitioners to monitor mechanical seal performance in real-time.

This lecture discusses the hardware and software required to implement a mechanical seal asset monitoring program. In addition, it discusses the theory behind a novel sensing method utilized in such technologies. The lecture underscores the benefits of monitoring these components in real-time. It includes case studies highlighting potential increases in seal and compressor reliability, and reductions in emissions from implementing such a program.

AUTHOR INDEX